'I have enjoyed an ex[...] passion and clarity h[...] nature of religious feeling in human life, demanding a quality of response from the reader that matches his own high standard of thought and exposition.'

Professor Chris Gosden,
School of Archaeology,
University of Oxford

'A look at the ultimately important questions of life that is itself wonderfully alive: you may not agree, but you will never be bored'.

Iain McGilchrist, Quondam Fellow of All Souls College,
Oxford and author of The Master and His Emissary:
The Divided Brain and the Making of the Western World

'In this tour of the weird and wacky in religion and spirituality, Charles Foster displays his gift for making science accessible and philosophy entertaining. He will amuse and irritate religious believers and non-believers in turn but won't let either group stray too far from the evidence. You may not agree with all of Foster's answers, but he is certainly asking good questions.'

Dr. Justin Barrett,
Director of the Institute of Cognitive
and Evolutionary Anthropology,
University of Oxford

Also by Charles Foster

The Selfless Gene

Wired for God?

The Biology of Spiritual Experience

CHARLES FOSTER

HODDER

Scripture quotations are from the New Revised Standard Version Bible,
copyright © 1989 National Council of the Churches of Christ
in the United States of America.
Used by permission. All rights reserved.

First published in Great Britain in 2010 by Hodder & Stoughton
An Hachette UK company
This paperback edition first published in 2011

2

A CIP catalogue record for this title is available from the British Library

ISBN 978 0 340 96443 9

Typeset in Monotype Sabon by Ellipsis Books Limited, Glasgow

Printed and bound by in the UK by CPI Mackays, Chatham ME5 8TD

Hodder & Stoughton policy is to use papers that are natural, renewable
and recyclable products and made from wood grown in sustainable forests.
The logging and manufacturing processes are expected to conform to
the environmental regulations of the country of origin.

Hodder & Stoughton Ltd
338 Euston Road
London NW1 3BH

www.hodder.co.uk

To Colin Roberts,
whose lessons in happy anarchy
on the moors of the Peak District
are responsible for a lot

If we can get back to our own world by jumping into *this* pool,
mightn't we get somewhere by jumping into one of the others?
Supposing there was a world at the bottom of every pool!

(C.S. Lewis, *The Magician's Nephew*, 1955)

Contents

List of Illustrations

List of Illustrations

Image Permissions

Images on the following pages are reproduced under the terms of a GNU Free Documentation Licence: 14, 17, 18, 26, 29, 52, 54, 62, 64, 75, 76, 90, 96, 100, 109, 110, 115, 140, 168, 175, 203, 214, 215, 228i, 228ii, 228iii, 230i, 230ii, 234.

Images on pages 98, and 212 and 213 are copyright Charles Foster.

I am grateful to the following for permission to reproduce images on the following pages:

12. Patrick Lynch: Creative Commons Attribution Sharealike.
20. and 254. Jolyon Troscianko (originally published in Susan Blackmore's *Consciousness: A Very Short Introduction* (2005): Oxford: OUP).
24. Istockphotos.com.
27. Michael Persinger.
32. C. J. Limb and A. R. Braun (2008) *Neural substrates of spontaneous musical performance: An MRI study of jazz improvisation:* PLoS ONE 3(2):e1679: doi.10.1371/journal.pone.0001679
38. Andrew Newberg.
60. The Bede Griffiths Collection, Graduate Theological Union Archives, Berkeley, CA: Copyright Ronald Ropers, 1991.
93, 94 and 95. Debra Varner.
114. Pablo Neo.

Acknowledgements

Many people have helped, lots of them without knowing it. As always, my debts are too many to list, but I have to mention:

Professor Chris Gosden, for guiding me through the Upper Palaeolithic;

Dr Justin Barrett, for helping me to make some sense of children's minds. He read the manuscript in draft and saved me from some deep embarrassment;

Dr Iain McGilchrist, who also read the manuscript in draft. His ground-breaking work on what it means to be a human being has long been an inspiration. And he lives it out too;

Professor Susan Blackmore, one of the bravest travellers on the wild frontiers of consciousness, with whom I disagree profoundly about lots, but probably less than she'd think;

Professor Simon Conway Morris, who does evolutionary biology with both his hemispheres;

Dr Anna Williams, a very clever neurologist;

Graham Hancock: honest, curious and humane in the face of extreme strangeness;

Dr Karl Jansen, whose work on the relationship between near-death experiences and experiences on ketamine opened some important doors;

Dr Michael Lloyd, who companionably shares my loathing of Gnosticism;

Dr Sally Hope, for trawling through my subconscious without audibly gagging;

'Monica', for extraordinary candour, and a lot of books on Tantric sex;

David Monteath, a fiercely kind critic;

All at Shantivanam Ashram, where I learned much, but not enough;

All at Holy Trinity Brompton, for everything;

My wife Mary, who puts up with a lot, and is fed up with reading acknowledgements of it in books, as if that makes it all right;

Wendy Grisham, Ruth Roff and all the amazing team at Hodder, and particularly Katherine Venn, my original editor at Hodder. This book was entirely her idea, and however the book has turned out, it was a great idea.

Preface

I am desperate to know if anything is real. I am desperate to know how to distinguish between the real and the bogus.

This book is an attempt to answer the question: 'What should we make of religious experience?' That is not at all the same question as: 'Can we locate religious experience to a particular part of the brain?' or: 'Do you always get a religious experience when you get a burst of serotonin?' Unfortunately the massive bibliographies of the subject are full of books which think that to localise an experience is to explain and expound it; that to describe a process is to give an account of its origins; that correlation is the same as causation. That makes for rather dull books (unless you think that *mere* anatomy or *mere* biochemistry is interesting) and very shallow theology. For anyone interested in the *significance* of religious experience, it is methodologically disastrous. If you start with a grey, reductionist premise, you shouldn't be surprised or disappointed when you get a grey, reductionist answer.

My starting premise is complete bafflement. I am baffled by the experiences I have had which I rather arbitrarily choose to label 'religious', and I am even more baffled by the inaccessibly odd experiences of Himalayan yogis, Sufis, levitators, alien-abductees, speakers-in-tongues, Peruvian shamans, takers of peyote, Presbyterian clergymen and German bankers.

Am I completely objective? Of course not: no one is. But genuine confusion probably approximates more closely to objectivity than do many other states of mind.

I would have preferred to avoid diving into the murky and intellectually shark-infested waters of the mind-body problem.

xvii

But it couldn't be done. There are two reasons. First, it is at least arguable that all religious experiences are steps along the road to realisation of Absolute Unitary Being (which many say is the goal of all religion). And you can't talk about that without talking about the problem of consciousness. And second, you can't talk intelligently about experience without knowing something about the experiencer. It would be like writing a book about the Russian Revolution without once mentioning Russia. But the chapter on consciousness is inevitably much more technical and inaccessible than the others. It has been relegated to an appendix, and can be skipped if you're feeling tired.

The book gets easier to read as it goes along. But you have to know the basic grammar of the subject (which includes learning some basic neurological wiring diagrams) before you can make sense of the really interesting texts. That grammar is introduced in Chapter 2.

I have used the ugly, lazy word 'God' to refer generally to the deity worshipped in the Abrahamic faiths. I wonder if I'm alone in finding the word a real disincentive to devotion? Names matter. It is hard to think nice things of someone or something with a name like that. Perhaps revival would be helped by reclaiming some of the beautiful Hebrew names such as Elohim or El Shaddai. And it is hard not to worship someone addressed as 'The Holy One, Blessed be He'.

Not much about religion is clear, but it is clear that Nietzsche's gloomy prognosis for religion was hopelessly wrong. The tenacity of religion perhaps indicates something about how fundamentally it is wired into us.

Charles Foster,
Oxford, June 2009

Prologue

On a Himalayan glacier a young man, naked except for a loin-cloth, sits cross-legged. His eyes are shut. His pulse and blood pressure are dramatically lower than normal. His breathing is almost undetectable.

In a London church a stockbroker, desperately worried about losing his job, is being prayed for. His eyes are shut too, but although he is fully awake, his eyes are in the rapid eye movement normally seen only in deep dream-sleep. Tears burst out. He falls to the floor, jerking and laughing.

It is dusk in the Amazonian basin. A group of Californian tourists waits nervously in a jungle hut. Vampire bats flap overhead. A deer coughs as it is killed by a jaguar. An old Peruvian man pours some murky fluid from a plastic lemonade bottle into some chipped cups. The Californians drink. They gag. Soon two of them get up to vomit. One of them sees his dead mother. Another strokes a snake as high as a house, which transmutes into an Egyptian goddess.

In a bedsit in Earls Court a couple are making love. 'O God, O God!' moans the girl. And she means it.

While sitting at home, watching TV, a middle-aged accountant feels crushing chest pains. He collapses. His wife, who has feared this for years, calls an ambulance. By the time it arrives the accountant is unconscious. He hovers above his own body, looking down with mild interest at the paramedics working frantically to restart his heart. He notes that the top of the door is dusty, but doesn't care much.

A mild-mannered man, whose main interests were running

the church youth club, playing with his children, and country walks with his adored wife, starts acting strangely. He gives up the youth club and takes up gambling. He gets into financial difficulties and spends on prostitutes the money that would have prevented the repossession of the house. He beats his wife and kills the children. While he is in prison an astute doctor asks for an MRI scan of his brain. The result is interesting.

In her cell in an Italian convent a nun is reading the Bible. She suddenly feels as if she is being transfixed by a thousand arrows. 'More of this pain,' she pleads.

An archaeologist shines his torch into a dark tunnel leading off a cave in south-west France. It has taken him an hour to crawl this far, and most of it was on his hands and knees. Some of it was on his belly. It would have been no different in the Upper Palaeolithic. On the wall of the tunnel are some beautiful paintings of bison. One of them has the head of a wolf. Next to them is a criss-cross pattern and some dots.

In a square in eastern Turkey dervishes begin their dance. With their arms outstretched, and in time to the thumping drum, they spin, getting faster and faster. They look solemn at the start, but soon they begin to smile.

An Indian Hindu and an American Episcopalian are badly injured in a bus crash in Mumbai. They both go along an identical tunnel towards a bright light and a reception committee of dead relatives. The Hindu sees Krishna. The Episcopalian sees Jesus. Both the Hindu and the Christian are told to go back to their hospital beds. They reluctantly obey.

In Kingston, Jamaica, a Rastafarian hands a huge spliff round the bar. 'Take it, man,' he says. 'It's the sacrament from God himself.'

'You're not far wrong,' says a Hasidic Jew who is sitting by the door, reading the Torah. 'The holy anointing oil in the Temple contained cannabis.'

'Does that mean', asks an earnest evangelical student from Nashville, not sure what his pastor would make of the spliff, 'that Jesus, the Anointed One, was actually anointed with marijuana?'

'You're finally getting the point,' says the Rastafarian, taking a deep drag.

What is going on?

CHAPTER I

Matter Matters:
Religious People are Made of Molecules too

A 9 year old boy is diagnosed with a benign brain tumor near his temporal lobe . . . [H]is parents and teachers notice a change in his personality. He becomes more aggressive, impulsive and extremely unpredictable. He growls at other students and occasionally hits them. He tells his parents that he can see visions and flashes of color. He also complains of smelling rotting meat . . . He is referred to a neurosurgeon, and the parents finally consent to removal of the tumor. 6 months later his pre-morbid personality, that of a quiet, friendly, sociable, loving young boy, returns.

(N. Ribner, *Handbook of Juvenile Forensic Psychology*, Hoboken, New Jersey: John Wiley, 2002)

'I'm just not interested in how things work,' said the vicar of a very conservative Protestant church, drinking the orange juice befitting a man who had to be up early next morning dispensing immutable truths to the multitudes. 'I'm not interested in cars, trains, or organisms. It doesn't matter to me how people are physically wired together. My business is with what they do with what they've got.'

I took a deep breath and moved on. I didn't know where to start.

This chapter is unambitious. Most people will not need to read it. It is directed only at those who believe that there is nothing that can happen to the material of our brains which can affect our ability to relate to God, if he is there. Many of the people who believe this will, by a deep and deeply unfortunate irony, be Christians.[1]

A clergyman who thinks that his job is the cure of souls needs to go back to theological college. He is paid to shepherd a flock of mind-body-spirit unities. If the Pope is clinically depressed, he will feel that God has left him. The world, once festooned with the trappings of divine favour, will be grey. If he can bring himself to open his Bible, he will find that the stories of God's action in history are unbelievable fairy tales, the assurances of God's loving concern empty platitudes, and the exhortations to mission exhortations to lie. But if he takes Prozac, he will probably feel better. Colour will stream back into the universe; he will believe again; relationship will be restored. If his depression is profound, he might need electro-convulsive therapy (ECT). Nobody knows how it works, but, rather like giving a broken television set a good kick, it often does. If he is bipolar, lithium will flatten out his psycho-rhythm. It will stop him being flung, Lucifer-like, from heaven to hell, but only by stopping him getting to psycho-heaven in the first place. The dark night of the soul won't be so dark, but the dazzling day of the Lord won't be so bright either.

The point is obvious: brain chemistry and the electrical environment of the brain affect our ability to feel religious things. Some extreme examples are discussed in this book (pills can induce visions that seem explicitly religious), but there are plenty of other mundane examples. If the vicar had drunk claret instead of orange juice, his congregation would have noticed. If he had drunk a bottle of whisky he would have had mood changes (probably involving intrusively violent and sexual thoughts), his ability to think his usual quiet sacred thoughts would have been compromised, and he would have fallen over. Eventually, when his liver enzymes had done their work, he would have been able to say his prayers again. If the cerebral cortex of a devout Christian healer is deprived of oxygen for twenty minutes, he will die. That will undoubtedly affect his ability to have spiritual experiences, but nobody has any real idea how.

In the 1950s and 1960s the physiologist Jose Delgado implanted electrodes into the brains of various animals and stimulated them. The results were spectacular. He could stop a charging bull by flicking a switch. We are bulls, and there is no difference in principle between stopping a charge and stopping a numinous experience. And if you can stop things, why should you not be able to start them? Michael Persinger thinks that he can generate a feeling of supernatural presence using his 'God helmet' to stimulate the temporal lobes. We look at his work later,[2] but the mainstream pathology books are full of examples which indicate that experiences we think of as spiritual or moral can be affected by damage to the brain. It has even been suggested that every mystical gift has a pathological corollary somewhere in the medical literature. Tell a neurologist about a beautiful picture that you have seen in your meditation session, and he will go to his shelf and show you a blasphemous parody of it, with EEG traces and MRI scans to boot.[3] Sometimes experiences and characteristics can be facilitated by the removal of the parts of the brain that normally inhibit them.

Remove the amygdala (as happens sometimes in epileptics), and your ability to experience negative emotions, such as fear, is massively impaired. Call it courage if you like. It is certainly tremendously dangerous. An amygdala-free brain is likely to find itself under a juggernaut or inside a lion. Tickle the amygdala electronically, and the stiffest upper lip will tremble as the world is populated by perceived horrors. A chair will terrify. Fear and awe are close relations: indeed perhaps the amygdala is the throne of awe. If that is so, perhaps incense ascends more directly to the throne of awe than the Old Testament writers ever imagined, for the very centre of the amygdala receives inputs from the olfactory system.[4] The use of incense in religious ritual is ubiquitous and neurobiologically comprehensible.

Religious people are typically very keen to say that we are all free to respond to God: that we are all entirely responsible for our own decisions. Really? Can the notion of free will survive, unedited, after a look at the literature?[5] What about the functional MRI studies of psychopaths that indicate a reduction in the normal control of the frontal lobes over the welling springs of emotion in the limbic system, and difficulty recognising and therefore relating to the emotional cues of others?[6] What about the violent behaviour associated with tumours of the limbic system? When the tumours are removed, so is the violence.[7] Autism (whose physical markers are increasingly recognised) smashes up the whole ability to relate to anyone and anything. Simply being male seems to be a form of autism: males generally have less of a Theory of Mind (TOM) than females,[8] and correspondingly less ability to sympathise and empathise.[9] Women are constitutionally nicer people than men, and seem to find it easier to believe in God.

Christians have no theology to deal with these things. When the Pope was depressed, was his faithlessness a sin? When children with limbic tumours swear at their parents and kick other children in the playground, do they need to be forgiven by God? When a sadistic psychopathic killer recalls, unrepentantly, and with the only delight of which he is capable, the agony of his victims, is he destined for the pit? Will men be judged less harshly than women because women start off in ethical pole position?

It is not just theologies that are inadequate: the law of most supposedly civilised countries has a similarly pre-scientific view of human transgression. While one can well understand the view that a psychopathic killer should be caged to prevent further offences, what justification can there be for the sentencing remarks that go with the life sentence, invoking a view of human autonomy drawn straight from John Stuart Mill?

The problems are not just problems associated with disease states. None of us is free. Our environment determines the experiences we have. You are many million times more likely to become an evangelical Christian if you were born in Alabama than if you were born in Riyadh. Your chances of speaking in tongues will be correspondingly much higher. If you are the son of a Virginian orthodontist you are less likely than the son of a Peruvian shaman to tussle with a malevolent parrot-spirit.

In this book we will come across, again and again, the vexed question of what is 'normal', of what we are *supposed* to see and hear. In our dreams we see all sorts of things: delights that we would be prosecuted for pursuing; horrors such as Dante could never imagine. Fortunately we are paralysed, and cannot act them out.[10] There are some people who cannot switch off their dreams, and cannot shut their ears to the ethereal voices that we all hear in our sleep. We call them psychotic, and try to drug their voices into silence. In another age they might have been called seers, prophets, or even gods.

CHAPTER 2

God Head: The Anatomy of Religion

The human brain does not contain a single 'God spot' responsible for mystical and religious experiences, a new study finds.
(*Live Science*, 29 August 2006)

Scientists locate 'God spot' in human brain.
(*Live Science*, 10 March 2009)

Trying to surmise the brain activation patterns of a cognitive task based on functional neuroimaging data may be like Noah trying to surmise the landscape of Mesopotamia after the Great Flood by staring at the peak of Mount Ararat protruding above the water.
(E. Goldberg, *The Executive Brain: Frontal Lobes and the Civilized Mind*, Oxford: Oxford University Press, 2001)

Richard Dawkins famously dismisses anyone who holds any religious belief as a 'faith head'. Like many of the things he says, it is half right. There is undoubtedly some correlation between some of the things that go on inside our brains and the experiences that we call 'religious'. It would be very surprising if there were not. And just think of the howls of Dawkinsian delight that would ring through the tabloids if there were no such correlation: 'Mindless faith', the headline would read. 'Professor Richard Dawkins, of Oxford University, told reporters that the PET scan findings confirmed what he had really thought all along. "Religion is not remotely an intellectual activity," he said.'

A word about correlation: it is not the same as causation. Being in the library correlates with the writing of this book; it does not cause it. There is a perfect correlation between being female and having ovarian cancer, but being female does not cause the cancer (although it is a necessary condition of it). Now think about neurological events. Imagine that I look at a dog. Visual information from the dog passes through my eye and my optic nerve to be processed in the brain. If you SPECT-scanned my brain while I was looking, you would see lots of activity in the visual processing areas. Has that activity *caused* the dog? Of course not, although it correlates perfectly with my dog-viewing. Does the fact of the observed brain activity mean that the dog is a delusion? Of course not. All this is almost too embarrassingly obvious to say, and yet it is often suggested, on analytically identical grounds, that we can confidently say that there is a 'God Delusion'.

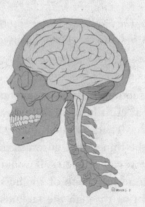

The human skull and brain.

So: real things go on in our brains when religious things happen to us. Not only that, but the things we do seem able to shape our surprisingly plastic brains. It can seriously be argued from the scientific literature that forcing young children to engage in religious practices, basketball or politics is physical child abuse in the same way as beating them with a cane – with the seriously aggravating difference that brains are often moulded for ever; buttocks are only bruised for a week.

The neurologist Vilanyanur Ramachandran, addressing the Society of Neuroscience in 1997, said, 'There is a neural basis for religious experience.' The observation was widely reported in both the lay and the scientific literature, and turns up tediously in most of the books on the subject, usually described as 'radical', 'shocking' or 'iconoclastic'. But why? Would we really expect it to be otherwise? Would we really expect things *not* to be going on in our heads when we are in ecstasy?

The idea that there is a 'God spot' in our brains is probably as old as the history of consciousness. Many Eastern religions talk about the 'third eye', thought to be located in the frontal cortex, and some practitioners went as far as to bore a small hole into the skull over the frontal cortex to increase the receptivity of the 'eye'.[1] Apparently it is catching on again. Amanda Fielding drilled a hole in her own forehead in 1970 in front of running movie cameras. The movie became a cult classic.[2] Many mystical religions have assumed that the pineal gland is the neurological palace of God. The 'uraeus snake', often seen rearing out of the forehead of ancient Egyptian priests, is sometimes claimed to be both a depiction of the pineal and an exposition of its function, and some say that the pine-cone topping the wands of Dionysiac devotees is the pineal.[3]

Every so often researchers excitedly claim that they have found that palace elsewhere. It is in the angular gyrus in the right cortex, says Professor Olaf Blanke, of Geneva University Hospital. It is in the medial temporal lobe, says Professor John Bradshaw of Monash University. The medial temporal lobe is rich in serotonin receptors, and serotonin is demonstrably

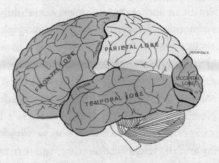

The outer surface of the brain.

involved in transcendental states. Well, not quite, says Michael Persinger, at Laurentian University. If you tweak your right temporal lobes electronically, the left hemisphere (the seat of language and reason) tries to make sense of the non-existent, electrically induced entity that appears to be affecting the right side, and misconstrues it as a real being. Label that being 'God', and you've got a theology. No, say researchers at the University of California, San Diego: the skull borers were on the right track – it is in the frontal cortex. 'But what about schizophrenics?' say others. 'They *hear* voices in apparently the same way that the great prophets did. Surely that means that the fictional God is somehow connected to the auditory processing system?' And so on. You can find someone to make a case for the God spot being in almost every part of the brain. Neurologically, God seems to be everywhere in my head. And that, in fact, is probably what the modern scientific consensus is. Neuroscientist Andrew Newberg says:

My work . . . strongly suggests that there is no God 'part' or 'module', but rather a complex network involving virtually the whole brain when these rich and diverse [religious] experiences are elicited. We can point to specific areas of the brain that may be associated with specific components of religious experiences, but since there are numerous ways to perceive, think about or meditate upon God, each method of meditation or prayer will affect the brain's function in slightly different ways.[4]

If no one spot in the brain is designated as a sacred space – an anatomical temple – is there a particular endogenous substance that turns on God-thoughts? You can certainly take God-pills: we deal with them in detail in Chapter 6. Perhaps prayer, meditation or worship works by causing the body to produce analogues of those pills? Perhaps certain people are religious because they have higher than normal levels of those analogues, or are particularly sensitive to them? Everyone knows the feeling of well-being that you get when you come back from strenuous exercise. Serotonin and dopamine have credibly been given the credit for that. Perhaps religious ecstasy is just what you get from your jog, mediated by serotonin but multiplied twenty times and translated into Latin, Hebrew, cloud-visions, orgasm, or any other religious language of your choice?

Well, maybe: there are few certainties in the neurology of religious experience, and today's heresy has an uncomfortable habit of becoming tomorrow's crusty orthodoxy. But there is no evidence for it. The American neurologist James Austin noted some animal studies showing that some naturally occurring chemicals can overstimulate and destroy certain types of neurone in certain parts of the brain. These chemicals are called 'excitotoxins'.[5] He suggested that mystical experiences can both cause and be caused by a release of natural excitotoxins. These compounds selectively destroy the neurones responsible for creating and storing anti-mystical feelings and attitudes. Continued religious practice, according to Austin, can literally mould your brain into a more religious shape.[6] At the moment most of us are amphibians: we can survive in the mystical and the non-mystical worlds, but are generally better at getting on in the non-mystical. But the longer we swim in numinous waters, the more acclimatised we get. Our mystical gills become physiologically more efficient. It's an interesting thesis, but entirely unsupported by any evidence.

Austin is keen, too, on the notion that naturally occurring opiates (morphine-like compounds) may be responsible for at least some mystical 'highs' – highs that may trigger a more consistent interest in, if not residence in, a mystical state. His keenness stems from a mountain-top high that he had when being given morphine for a surgical procedure. It was an experience so extraordinarily sweet that he has been scared stiff of morphine ever since.[7]

On the face of it, this is credible. The action of endogenous opiates is well known: just like their cousins from the pharmacy or the street corner, they produce bliss, peace, sometimes a sense of disembodiment, a reduced respiratory rate and a reduced heart rate. Just about the whole suite of effects, in fact, that you will record if you wire up a world-class yogi to monitoring equipment and ask him to describe his experience of *kensho*. But, again, it doesn't square with the evidence. If you block the action of opiates using naloxone, you don't seem to block the meditator's *kensho*.[8]

If you are trying to run God to ground in the brain, probably a more promising place to look is not amongst individual compounds, nor in cocktails of compounds, nor in particular spots, but in the relationship between the two hemispheres.

The popular understanding of the role of the two hemispheres is broadly correct.[9] The right hemisphere is the feminine, intuitive hemisphere. It is a device for painting, or at least feeling, the big picture. It's the temple of impression, sympathy and holistic understanding. It questions the adequacy of the models of the world drafted by the left hemisphere, and is capable of exhilarating iconoclasm. If your worldview is to be ripped up, the right hemisphere will do it. It is not good at footnotes – it leaves those to the left – but if the number of caveats to be inserted into an idea becomes unsustainably heavy, the right hemisphere will become suspicious of the validity of the idea. The right hemisphere is far better company than the left. At dinner you'd want to sit next to the right hemisphere, not the left.

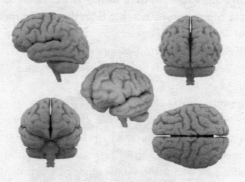

Five views of the human brain, showing clearly the division between the hemispheres, bridged by the corpus callosum.

The left hemisphere is achingly reductionist. It squeezes all the intuitions of the right into conceptual pigeonholes. It is very keen on words and subclauses. It translates feelings into more or less coherent (or at least describable) systems. It dissects, categorises and files. It wears a suit and rarely smiles. It is terrifyingly conservative, and fights tooth and nail with the right side to preserve its precious models. It is horrified when the cavalier right hemisphere suggests that a model has become superannuated. If a carefully constructed system seems inadequate, it will, just as Freud suggested, lie, suppress and turn its Nelsonian blind eye to the inadequacy. Happy are those who have learned to surrender to the rule of the right.[10] And surrender can be learned. Yet the right without the left is amorphous or anarchic. Whole people are rational mystics, intelligent ecstatics. The most sublime poetry is not a stream of random, orgiastic ravings. It seems, fascinatingly, that women might be literally better integrated people than men. The primary road between the hemispheres is a bundle of neurons called the corpus callosum. Women have a thicker corpus callosum than men: the left brain and the right brain talk better to one another than they do in men.

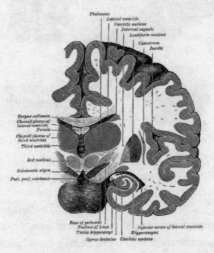

Side to side section of the brain, showing the location of the thalamus and corpus callosum.

The whole history of humankind and every individual within humankind can be written in terms of the battle between the hemispheres.

I have personalised the hemispheres in this description. Anthropomorphism is difficult to avoid. We have heard already that some say that the whole idea of a 'sensed presence' – so prominent in much of religious experience – is a consequence of one hemisphere anthropomorphically seeing the other as a real entity. And indeed there are circumstances in which the hemispheres act more or less separately. Those circumstances provide good evidence for the suggestion that each hemisphere can have a consciousness of its own. And if that is so, might not our gods be merely one or other half of ourselves? Might Jesus not have been telling the literal neurological truth when he insisted that 'I and my Father are one'?

Epilepsy is an electrical storm in the brain. From the primary focus – the eye of the storm – wild electric gusts blow through the brain. They can cause massive disruption of normal brain function. It is possible to stop the storm crossing from one hemisphere to the other by cutting the corpus callosum. On the surface, patients who have had this procedure seem pretty normal. But if you start probing, you see some very strange things.

To understand why, we need to know a bit about the brain's wiring. Information from the right visual field (even in these patients with a divided corpus callosum) passes to the left hemisphere, and vice versa. Language, very, very roughly, is dealt with by the left hemisphere. If you show a picture of a cat to the right visual field of a split-brain patient, the image will pass to the left hemisphere, and when the patient is asked to describe what they have seen, there will be no problem. 'A cat,' they will say. But if the image is shown to the left visual field, the image passes to the right hemisphere, which basically doesn't do language. Then, if asked to explain what has been seen, the patient will not be able to describe it in words, although they may be able somehow to indicate it using the left hand (because the left hand is controlled by the right hemisphere).

The plot thickens. In a famous experiment, a snow scene was shown to the left visual field of a split-brain patient.[11] A chicken

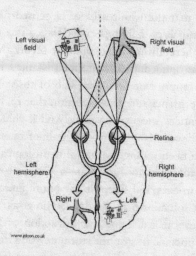

Left visual field

Right visual field

Retina

Left hemisphere

Right hemisphere

Right

Left

www.jolyon.co.uk

An experiment on a 'split-brain' patient, indicating that to some extent the hemispheres can operate independently of one another. (Jolyon Troscianko)

claw was shown to the right visual field. The patient was then asked to pick out matching pictures. The left hand picked up a picture of a shovel – an entirely rational thing to do: the right hemisphere was thinking of clearing snow. The left hand picked up a picture of a chicken: again a sensible connection. But when the patient was asked to explain his reasons for picking the pictures, something weird happened. A verbal explanation could of course be generated only by the left brain, which knew nothing about the activities of the right brain. The left brain was articulate, creative and utterly dishonest. It seems to have been embarrassed by its ignorance, and it covered up its shame by an imaginative confabulation: 'Oh, that's simple: the chicken claw goes with the chicken, and you need a shovel to clean out the chicken shed.'

This is fascinating and disturbing. How many entities lived inside that head? Were there two conscious entities, one on each side, one more verbally articulate than the other?[12] Was the 'real' conscious person only and ever a resident of the left side?[13] Had the operation simply made the real 'self' of the patient migrate to, and take up permanent, exclusive residence in, the left hemisphere? The patient wasn't 'consciously' confabulating ('Which consciousness?' you might well ask). The connection between the chicken claw and the shovel was an honest expression of what 'he' thought. Surely that should make us pause when we are insisting to ourselves and others that we are giving our honest opinion about something. If the right brain could have spoken, it would no doubt have given a different interpretation of the choice of the images. If we can't be sure even about what we think we're thinking, can we be sure of anything at all?

We can multiply entertainingly the difficulties posed by these split-brain patients. Suppose you're the patient's vicar. You will know that for some of the time the patient will be lying when he says to you that he thinks X or Y. Do you rebuke him for his soul-endangering dishonesty? Suppose that the patient, when wearing blinkers that obstruct one visual field, goes to see an obviously corrupting pornographic movie. Do you rebuke just the side to which the movie has projected? Or both sides? Or if the movie has projected only to the right side, do you assume that there is no self there capable of depravity? How many souls does the patient have? Can the right hemisphere be 'saved' by an appropriate response to an evangelistic sermon without dragging the left hemisphere with it through the pearly gates? And so on.

This is all good fun, but it is also desperately, fundamentally serious. It goes to the root of who and what we are. Some of the threads are picked up in the Appendix, which deals in detail with the problem of consciousness, but for the moment it is enough to state the questions that split-brain patients raise, noting that these patients represent the teetering heights of the reductionists' case – the case that when we feel 'god' we are simply feeling the presence of another 'consciousness' just across the corpus callosum from the receiving hemisphere.

In split-brain patients, they say, there is none of the confusion between correlation and causation about which I lectured so portentously. The only conceivable explanation for the chicken claw confabulation is the divided brain: ordinary people don't confabulate that way.

Fair enough. But let's not push things too far. Split-brain patients are so exotic, and the experiments so gratifyingly weird, that it is hard not to get overexcited. But do they really tell us anything that we didn't know already? Hardly: we opened the discussion of the role of the hemispheres by admitting that they had very different characteristics, different views of the world, and spent their time in exhausting discourse with one another. Say that they have different consciousnesses if it pleases you: it doesn't really help you to elbow God out of the picture. You might even speculate that, precisely because they are in such intimate commerce, they know one another so well that they are unlikely to mistake each other for God. One can just about understand a South Sea islander ascribing divinity to a European dressed in military uniform and landing from a battleship.[14] But my wife, who knows my inadequacies and washes my socks, is unlikely to make the same mistake about me. The right and left hemispheres are more like Darby and Joan than the Cargo cultee and the swagger-stick wielder. As jihad, the battle of the hemispheres is wholly inconclusive.[15] For, as we will see, not only can the reductionists not use anything in modern neuroscience to elbow God out of the picture, there's little there that the faith heads can use to shoehorn him in.

Our brains are colossal: far, far bigger in proportion to our body weights than any other organism on the planet. Although there is no very obvious correlation between the size of someone's brain and their accomplishment, sheer brain size presumably does confer some sort of adaptive advantage. We pay a high metabolic cost for our big brains: they are greedy energy-consumers. And also they are very dangerous. Squeezing out the head of a human child is far riskier for both mother and foetus than squeezing out the head of a calf. And even the birth size of a human head is a compromise between size and maturity. Human infants, so that their heads are still small enough to be pushed through the birth canal, are born life-endangeringly premature. They can't run from predators or look after themselves for years, unlike the young of many other mammals, and make horrific demands on their parents in terms of energy, time and school fees.

Yet although there is presumably some biological point in being as big-headed as we are, the business of neurological development in post-term humans is a process of neuronal *loss*. You have more neurones as a baby than you do as an adult. In young children there is a particularly luxuriant efflorescence of neurones in the frontal lobes – regions known to be associated with creativity. And it shows. Children are immensely imaginative. Pejoratively, we tend to label the creatures of their imagination as lies. If you show a young child a picture and ask him to describe it, the description is likely to be populated with characters and objects that are not there. Or at least that we do not see there. To a four-year-old a grey industrial landscape, yawned through by us, crawls with elves and dragons. As the child grows up the elves die off. By the time the child is twelve the landscape is about as dull for them as it is for us. We call it 'maturity', 'seeing things as they really are', or 'telling the truth'. But what do we mean by the 'truth'?

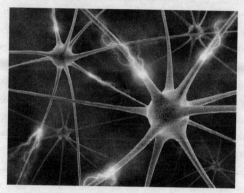

Electrically fizzing, intimately interconnected neurones.

There are few uncontentious things in neuroscience, but here is one of them. What we perceive as 'reality' is nothing of the kind.[16] Our sense receptors receive massively more information than they can possibly process properly, and so our brains are highly selective in what they use. Aldous Huxley spoke of the brain as a sort of 'reducing valve', and he was right. The function of the brain, nervous system and sense organs is 'in the main *eliminative* and not productive', he wrote. It is a filter that stops us (the conservative might say *protects* us)

> from being overwhelmed by [a] mass of useless and irrelevant knowledge, by shutting out most of what we should otherwise perceive or remember at any moment, and leaving only that very small and special selection which is likely to be practically useful . . . What comes out at the other end is a measly trickle of the kind of consciousness which will help us to stay alive on the surface of this particular planet . . . Most people, most of the time, know only what comes through the reducing valve and is consecrated as genuinely real by local language.[17]

But not all the time. And there are some people for whom this is never true. Those other times and those other people feature very prominently in this book. But let's pause for a moment on the thought that children generally – not the born shaman – may see more than we do; may have a view of the world based on a greater amount of data than we do. They seem to have more raw processing power.[18] As a brain grows older, new inter-neuronal connections form. The wiring becomes more like the wiring of other brains. There's less variety amongst adults than amongst children. And that's a shame. Tendencies to conformism rarely produce really interesting things. As brains grow up, the reducing valve tightens. Less is allowed to come in, and the product is an increasingly stagnant dribble.

What might we be if the valve were slackened off? What were we once? Are we really richer than we were? When Wordsworth wrote 'Intimations of Immortality from Recollections of Early Childhood', he was hardly celebrating the tightening of the valve. What are the neurological corollaries of Jesus' insistence that if you want to enter the kingdom of heaven you have to come as a little child?[19] If Freud is right, and our behaviour is driven by a desperate desire to get back to our childhood, why should that be? Surely we are built for reality, and therefore thirst for it. Perhaps, just perhaps, our colossal infant brains, getting more information than we do, come to conclusions about the world that are more reliable than our own. Perhaps the industrial landscape is more exciting than we think. As we will see, if we take children's instincts about God seriously, we're not far at all from a pretty orthodox theistic philosophy.[20]

We have seen already that a tumour somewhere in your hippocampus or a high-speed journey through a car windscreen can have profound effects on your personality and your ability to enjoy evensong. We will see that dehydration, dancing, pain, sensory deprivation and many substances can usher us into other layers of being. But strange, more subtle effects of long-term meditation practice are being uncovered by some new imaging technology.

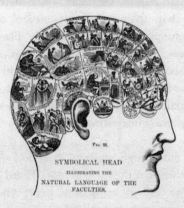

A phrenological diagram of the human head, attributing characteristics to various areas. Much modern neurology is discussed in frankly phrenological terms. (Samuel Wells, 1878)

Andrew Newberg used Single Photon Emission CT ('SPECT') scanning to look at what happened to the brains of Buddhist monks and Christian nuns.[21] These were very experienced practitioners. They had all been meditating very regularly and very intensely for many years.

The monks meditated for about an hour, focusing on a sacred object. The nuns used the 'centring prayer', first described in the fourteenth-century text *The Cloud of Unknowing*. Don't seek God through knowledge, the anonymous monastic author urges. That path is built on presumption: it is no way to the throne. Seek him instead with 'naked intent'. Only then will you be clothed with glory. Seek him with blind love. Only then will you see. Taste in order to see.

When the meditators reached the summit of their experience – a destination they knew well – they signalled to the investigators, who immediately injected into a blood vessel, via a preplaced catheter, a radioactive tracer. The activity of brain cells correlates well with the blood supply to them, and the amount of radioactive tracer passing to the brain in turn correlates with the blood supply. Thus when the brains of the meditators were scanned very shortly after they had reached the mountain top, the pictures gave a fairly reliable indication of which parts of the brain were active at the climactic moment. Non-meditating baseline scans had previously been done.

Michael Persinger, of Laurentian University, Canada, who claims that he can induce religious-type experience using his 'God helmet' See pages 54–57.

Some of the results were entirely predictable. Since the nuns were concentrating on words in the prayer, they showed much higher activity in the language processing centres than did the Buddhists. The Buddhists, correspondingly, had very active visual areas. And then it got interesting. Both the monks and the nuns had said that the climax included a feeling of timelessness and spacelessness; they felt that the boundaries of their respective selves had dissolved, or at least softened; they felt an intimate communion with the universe – as if they had assimilated into it; they felt a concomitant loving solidarity with everything that there was. They felt that they had escaped the tedious constraints of Newtonian space and time, and were floating in eternity. If this was not arrival at the supreme goal of the mystic – 'Absolute Unitary Being' (absolute identity with everything else in the universe, and experiential knowledge of the fallacy of dualism) – and Newberg concedes that it was not, it was arrival in its forecourts.

The nuns and the monks both had a marked reduction in the activity of the parietal lobes. The parietal lobes are responsible for telling the body where its component parts are in space.[22] They draw the frontiers of what belongs to us. If you have a tumour in your right parietal lobe, you might find yourself disowning your own leg. You might try to throw the alien, invading leg out of bed.[23] In an extreme case you might even disown your own reflection. Newberg relates reduced activity in the parietal lobes to a reduced sense of the discrete nature of the meditators' selves. And surely he is right. They didn't know where they ended and where the outside world began.

There is more. The meditators all said that the state to which their meditation had taken them was more real than the humdrum world of the cloister, the cell and the shopping mall. Why might this be? One explanation is that it *was* more real. St Paul observed that, we see now 'through a glass, darkly'. Perhaps in ecstatic states we see the ground of reality 'face to face'.[24] But there is another. Or perhaps it is not another at all. Perhaps it simply correlates with a real audience with the Ultimate, rather than being the cause of an apparent but illusory audience.

Deep in the centre of the brain lies a hazelnut-sized structure called the thalamus. It acts as a sort of sluice gate, regulating the flow of sensory data to many parts of the brain. If the flood gate is kept wide open, with lots of information racing through it, the higher centres of the brain – notably in the frontal lobes – will be kept busy, and will say to themselves and the rest of the brain, 'Business as normal at the moment. This is a classic Monday morning with a mountainous in-tray. Nothing odd or numinous going on here.' If the sluice gate is particularly wide, the higher centres may well say, 'This is particularly busy. This is really intense. This is life in all its fullness.'

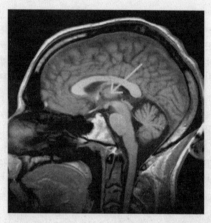

An MRI scan of the human brain, showing (arrowed) the thalamus – a sort of neurological sluice gate that regulates the flow of information to other parts of the brain. Thalamic activity changes strangely during some forms of religious activity.

Normally, when a subject is awake, the activities of the thalamus and the parietal lobes go up and down together. This is what you'd expect. As more information gets beamed into the brain, there is more filtering work for the thalamus to do, and the parietal lobes will be correspondingly busy defining where the boundaries of the body are in the light of the new information. But during the nuns' and the monks' meditation, the activities of the thalamus and the parietal lobes were *inversely* related. That's odd. Nobody has any idea why or how it happens, but one possible consequence is this: just as the parietal lobes are experiencing a strange state for a waking person – disorientation and blurred boundaries of self – the higher centres are saying, 'This is real: in fact it is *more* real than usual.'[25]

Newberg tiptoes cautiously around the causation/correlation trap:

> For all we know, the thalamus could be responding to incoming stimuli from an unrecognised or unseen source (which some people might call God), but it could also be responding to the conceptual activity that is occurring in various parts of the brain.[26]

The work on the neural basis of religious experience is just beginning. Newberg (who is by no means the only active researcher in the field)[27] suggests that many so-called mystical states might be associated with unusual relationships between the two components of the autonomic nervous system – the sympathetic and parasympathetic systems. Very roughly, the sympathetic system normally gears us up for action. It squirts adrenaline into our bloodstream, raises our heart rate, pumps blood into our muscles and increases our alertness. It is the 'fight or flight' system, triggered by coffee, fear, short skirts and ample lunchboxes. The parasympathetic system calms us down. Ideally you want your parasympathetic to rule at bedtime. Its bid for power can be helped by slippers, hot milk, a warm dog and a dull book. Normally, and obviously, the sympathetic and the parasympathetic systems are antagonists. But not always. And when they are not, some strange things can happen.

First, though, the near total rule of either of the systems can produce some interesting states. The sovereign parasympathetic can produce feelings of blissful tranquillity, as if you are floating, unconnected to the ties of the body and the demands of the outside world. It is hard to get there when you are awake, but it is found in some profound meditative states. Chanting, gentle drumming or a pitch-dark flotation tank might help.[28] The sales of Gregorian chant CDs and aromatic bath gel rocket in a recession.

The ecstatic dances of the Sufis, the San bushmen, any amphetamine-fuelled teenage clubber and the wild drumming that usually goes along with them can hand power to the sympathetic. In the ecstasy the dancers seem invincible. They are driven by a power that seems to come from outside themselves – an energy channelled directly into and through their consciousness. They become a river of power.[29] The river sweeps away all doubt and all weakness.

Sometimes, though, the psyche, or something, seems to broker a paradoxical peace between the sympathetic and the parasympathetic. In the moments of ceasefire, God appears.

Where the parasympathetic system is massively aroused, the calm can be so colossal that the body seems to forget for a moment that there is a war on. The mechanisms that normally inhibit the sympathetic are lulled to sleep.[30] And without any brakes on, the sympathetic explodes into life. The body lives the paradox of being at the same time profoundly awake and profoundly asleep. The result is an altered state of consciousness in which the entirely counter-intuitive becomes the norm. Paradox is a fine instrument for surveying the queasy no-man's-land between sleep and waking, and possibly between life and death too. The parasympathetic bliss coexists with the sympathetic energy. It is not far from the saints' descriptions of the beatific vision.

A similar thing can happen when, in a state of maximal sympathetic arousal (perhaps briefly during sexual orgasm, or in the throes of a stamping bush dance in the Kalahari, where all the available adrenaline is surging through the bloodstream to keep the dancer on his feet for his twentieth dehydrated hour), the parasympathetic bursts through, perhaps to insist, in its loudest voice, 'Enough.' The sympathetic energy is flooded with the colour of parasympathetic bliss. Perhaps the dancer is not far from wherever the Seventh Heaven is.

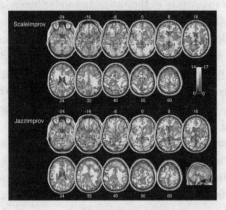

The biology of creativity. Functional MRI images of a pianist's brain when doing scales (top) and jazz improvisation (bottom).

Many will covet these sorts of experiences, and will want to get there without the help of psilocybin mushrooms or ayahuasca. Can they?

The answer is: possibly, but even if it is possible, the road is long and arduous. The monks and nuns in Newberg's and D'Aquili's SPECT studies had served strenuous apprenticeships in the night time of their souls and the dark of their chapels and meditation halls. There is a curious suggestion in the literature that these super-meditators might have brains rather different from 'normal' people. The inverted commas are important. The super-meditators might well be the intended norm, and we shallow, monochrome people the aberrations. In any event, Newberg and D'Aquili found that the *resting*, non-meditating SPECT scans of both the nuns and the monks showed asymmetric activity in the thalamus of a degree usually seen in serious pathological conditions.[31]

The significance, if any, of this asymmetry is obscure. But there are some interesting speculations. Conditions such as depression, dementia and Obsessive Compulsive Disorder (OCD) are often associated with disturbances in thalamic function, and the normal thalamic equilibrium can be disturbed by epileptic seizures.[32] Newberg speculates that it is the imprimatur of the disturbed thalamus that lends a sense of reality to the visions seen by epileptics – visions which often have explicitly religious connotations[33, 34] – and that the thalamic asymmetry of the prodigious meditators gives them a unique perception of reality even when they are eating, chatting or running for a bus. It suffuses the workaday world with glory.

We have no idea at all whether the monks and nuns were born with these thalamic asymmetries, or whether one side of the thalamus has ballooned out with the constant practice of meditation, just as your biceps will hypertrophy if you spend enough time in the gym. But we can say that the asymmetries are not pathological. These meditators are not physically, psychiatrically or psychologically inadequate. They have a facility that we don't have, or don't have easily. If access into these blissful, nirvanic states is really a result of hard training, there is a sound neurobiological basis for sermons about steep and narrow roads.

Whether the ability to enter these states is congenital or acquired, it is clear that we don't leave our pasts behind when we go into wherever it is we go when we pray or meditate. The nuns interpreted their experiences in the light of reality. But they wouldn't be able to put it into words, because language is a highly interpretative process. Their theology told them that

33

they had been in the presence of God. The Buddhists had tasted absolute consciousness.[35] It seems likely to Newberg, and I agree with him, that the preconceptions with which we come to the meditation hall restrict the palette that can be used to paint the divine landscape:

> I suspect that if a person could maintain a more open-minded state, the range of interpretations concerning spiritual experience might increase . . . If practitioners could meditate to suspend the brain's propensity to make interpretations, they might glimpse a truer reality.[36]

So, even in transcendental ecstasy the reducing valve might be stopping something important coming through. But this time the reducing valve might be prevented from opening completely by our past prejudices. It might be more fun, and more spiritually productive, to be a liberal.

So, is there such a thing as a naturally religious or naturally irreligious person? If there is some sort of biological predestination, all theologies of mission would have to be ripped up. We look in another chapter (and very sceptically) at the notion that there is a 'God gene'.[37] Before we get there, there are several things to say. The first is a warning that applies to most of the things said in this book. We return to it right at the end. This book is about spiritual experience: it is not about the very many other components of the religious life. Many people who would describe themselves as religious have never had anything like the experiences examined by Newberg and his colleagues. I think they are the poorer for it, and that they should crave such things, but then who am I? The second point is that the evidence is patchy, inconclusive and open to many interpretations. But the third is that, uncomfortable though it may be to our cherished Western ideas of free will, brain architecture and biochemistry do seem to confer real biological biases towards and against belief. These are only tendencies, though. Although it may genuinely be difficult for Richard Dawkins to feel religious sentiment, he is not neurologically predestined to damnation. No one is a hopeless case.

The evidence for some bias is clear enough, and surely accords with the experience of every evangelist from every creed. In a study in Zurich, scrambled words and phrases were projected onto a screen. Religious believers were much more likely than unbelievers to construct meanings out of the chaos. But when real words and faces were projected amid the jumble, the believers were significantly more likely than the unbelievers to see them. If one asked at the end of this phase of the study, 'Who is a better perceiver of reality?', honours would be pretty even. The study went on to look at what happened when the subjects were given L-Dopa – a precursor of the naturally synthesised compound dopamine, which is (and not just by this research) implicated in spiritual experience. L-Dopa increased the sceptics' tendency to conjure real words and images when they were not there.[38]

Studies of the resting brains of intellectual sceptics show increased frontal lobe activity and reduced activity in the hippocampus and the right caudate nucleus – results that suggest a dominance of the analytical over the emotional brain.[39] It has also been suggested that the balance of activity between the left and right hemispheres might affect religious or sceptical orientation.[40] Right-brained people will tend to be more religious than left-brained people.[41] This is an unsurprising result. Cynics will say that the study was pointless, and that all you needed to do was to count up the relative numbers of women and men in any church in the world. But the conclusion you draw from the data about the truth or otherwise of religious apprehensions will depend on whether you think that a holistic, right-brain sweep of reality is more likely to net God than a left-brained, language-based search which will only register a catch if the catch happens to fit into one of its tight, carefully crafted categories.

All this talk about biological biases might make us wonder whether true 'conversion' is possible. Newberg thinks that real, sudden conversions in owners of normal, inelastic adult brains are extremely rare.

> In a sudden conversion – when a person shifts from disbelief to belief, or shifts from belief to disbelief – the brain itself would also have to undergo a biological transformation in order to accommodate the new outlook . . . Then the question of how, why and where transformative experiences occur in the brain – and whether they make permanent changes in neural activity – remains a scientific mystery.[42]

But isn't this pretty meaningless? Conversion from what to what? The only really significant conversion would be from total disbelief in the possibility of the numinous to a belief in the numinous. I've never met anyone with such total disbelief. I doubt it is biologically possible, for reasons to which we will come. Most other 'conversions' are surely a more or less perceptible hardening of convictions and/or a renaming of convictions already held. An amorphous sense of joy is relabelled 'God', for instance, or, in more dramatic instances, 'Yahweh' is renamed 'Krishna'. But a relabelling of convictions is unlikely to demand a wholescale revision of brain biology.

So much for the evidence of the scans. They are interesting, but inconclusive. They do not tell us whether believers are born or made, although they do suggest that, however their brains got that way, believers might find it constitutionally easier to believe than non-believers. They tend to suggest that there is no one 'God spot' in the brain, but that when we have religious experiences many different parts of the brain contribute to them. The experiences are themselves coloured by our theology. The scans and the other data, by and large, do not begin to tell us anything at all about whether the experiences are self-generated or whether the brain is actually experiencing something other than itself.[43] The footmarks of God do not appear on the SPECT scans.

There is one possible and very eerie exception.

The phenomenon of *glossolalia* – 'speaking in tongues' – is a well-known feature of charismatic Christianity. It seems to have been common in the early church. St Paul seems to have expected believers to speak in tongues, and exhorted those who did not to ask for the gift.[44] It is not exclusive to Christianity, but is practised by the shamans of many tribes.[45] It involves the practitioner speaking in a non-human language. Usually the speaker will have no idea what the words are intended to mean. Sometimes the utterances will be (purportedly) interpreted by another. Charismatic Christians may talk about *glossolalia* as 'the language of the angels', or as 'the Holy Spirit speaking'. 'Tongues' are often used as a form of prayer when human words fail; when the subject is too big, or too emotionally charged. The words uttered are typically perfectly articulate, although particular phrases may be repeated over and over again. They sound like an ordinary but unknown language. When speaking in tongues the practitioner will not apparently be in any kind of trance. Someone who has the 'gift' will be able to start and to stop speaking in tongues at any time.

Work in the 1980s at Carleton University concluded that tongues were a learned pseudo-language. Non-tongues-speakers, after some coaching, managed to ape the sounds convincingly.[46] It was a shoddy study, and Newberg, using the SPECT technology to image the brain of a Pentecostal tongues-speaker, came to a very

American neuroscientist Andrew Newberg, who has used
SPECT studies to look at brain activity during various types
of spiritual experience.

different conclusion. There were some predictable findings, such as a reduction in frontal lobe activity: one wouldn't expect the users of an ecstatic mystical language to be engaged in exhausting analysis. But one would expect the language centres to light up brightly. Highly articulated words were emerging.[47] *But it didn't happen.* The language pathways seemed to be bypassed entirely. It was as if the words were being beamed directly to the tongue from outside. Newberg was excited, baffled and cautious. There was, and is, no known neurobiological explanation:[48]

This is a very unusual finding, for it suggests that the language was being generated in a different way, or possibly from some place other than the normal processing centers of speech. For believers, this experience could be taken as proof that another 'entity' had actually spoken things through them. For disbelievers, it might simply mean that other unique circuits, possibly associated with the thalamus, which directed the style and form of glossolalic speech, were being stimulated.[49]

'With the field of neurotheology expanding rapidly,' says Victoria Powell, 'it is possible that science will kill off the deity once and for all.'[50] Well, possibly. But assertions of his death are often made on scientifically very dubious grounds. God has outlived Nietzsche, who announced his demise. I dare say he will outlive Dawkins and Newberg too. It is really very difficult to imagine the instruments that any theopathologist would need in order to pronounce death with confidence. There is at the moment no way of saying with any scientific coherence that 'God' is the creature of the electrical tempests in our heads, rather than their creator. Indeed, the existing evidence is perhaps very marginally the other way. If you want to remain an undisturbed reductionist, you would be well advised to steer clear of the SPECT lab.

CHAPTER 3

The Holy Helix:
Genetically Predestined to Believe?

I call heaven and earth to witness against you today that I
have set before you life and death, blessings and curses. Choose
life. . .

(Deuteronomy 30:19)

When the Gentiles heard this, they were glad and praised the
word of the Lord; and as many as had been destined for eternal
life became believers.

(Acts 13:48)

That genes should have a large effect on personality, behaviour
and our choice of company should come as a relief to anyone
who values individual liberty. It means we cannot always be
conditioned by education or society to act as others want. The
Jesuit maxim – 'Give me the child for his first seven years, and
I'll give you the man' – does not always hold. Our genes allow
us to be ourselves.

(Mark Henderson, *The Times*, 11 August 2007)

'I'm just not the religious sort,' said my brother-in-law, Mick.
'Never have been, never will be. Churches bore me; Buddhist
chanting sends me to sleep; religious art of all kinds leaves me
cold; I'd rather read a cereal packet than the Bible; I'm mildly
amused when someone bellows at me that I'm eternally doomed;
I'm not hugely interested in what will happen when my heart
finally stops. Why am I here? I don't know, I don't much care,

and I'm happy. I don't wake up in the early hours full of existential angst. I don't want to read your books, and don't understand why anyone else would want to either. So drink up, and leave me alone.'

Evangelists want us to believe that Mick is not telling the truth: that deep down he, and everyone, is yearning to understand and to experience; that if their particular version of the truth is presented attractively Mick will wake up and say, slapping his ample forehead, 'Of course. This is what I've been waiting for all my life.' But it doesn't seem that way with Mick.

Nor does it seem that way for many people. The consistent experience of would-be proselytisers is that there are some people who get it, and some people who don't. The Calvinistic doctrine of predestination fits the facts very well. The people who don't get it aren't obviously depraved. Some of them are the nicest people I know. They don't wander through life with their metaphorical knuckles trailing along the ground, fornicating frantically, exploiting cynically and plotting the deaths of millions. In fact, for big-time evil, you can look far more reliably to the religions. Secular humanists don't fly planes into tower blocks. If it's a carapace of sin that is stopping God getting through to Mick, it's odd that the apparently much thicker carapace on the back of many of the world's fanatically religious figures doesn't seem to have kept him out of them.

It is becoming conventional to blame everything on one's genes. Can Mick point the finger at his DNA and say, 'You made me godless'?

In his 2004 book, *The God Gene: How Faith is Hardwired into our Genes*,[1] the geneticist Dean Hamer, Director of the Gene Structure and Regulation Unit at the US National Cancer Institute, implied that he could. You could quantify spirituality psychometrically, said Hamer, show that the tendency to this measured spirituality is partially heritable and, most controversially, show that part of the heritability is attributable to the gene VMAT2.

VMAT2 codes for a monoamine transporter that is central in regulating brain levels of dopamine, noradrenaline and serotonin. Those compounds may well be involved in the modulation of religious experience. Perhaps Mick is incurably irreligious because his genetically determined monoamine levels don't push his neurones into mysticism.

What would Darwin say about all this? It's not clear. Hamer has suggested, rather lamely, that the self-transcendence so emphatically denied to Mick increases optimism, which in turn increases well-being and so the chance of reproduction. Or perhaps self-transcendence is an incidental by-product of some other, more fundamental characteristic determined by VMAT2.

How do you measure 'spirituality'? Can we be sure that Mick is telling the truth about how constitutionally irreligious he is? Might a monk be hiding deep in his psyche? Indeed we can be sure, says the psychologist Robert Cloninger. If there is a monk, he will be discovered.

There are three elements to spirituality, according to Cloninger, each of which can be weighed using pretty standard psychometric principles of the sort deployed in every HR department. First there is 'mysticism', characterised by a readiness to believe things that cannot be formally proved. Then there is 'transpersonal identification' – a sense that one is connected to a universe of which one is only a part. And finally there is 'self-forgetfulness', manifested, for instance, by a tendency to become absorbed in some activity to the exclusion of everything else. I think that Mick would score pretty highly in each of those categories.

Twin studies, though, are bad news for Mick and good news for orthodox Calvinists.[2] They have shown that high spirituality scores are heritable. Hamer's contribution was to note that the spirituality index correlates strongly with the VMAT2 gene. We shouldn't overstate this conclusion, although Hamer has sometimes seemed to do so. He has sometimes asserted it with a confidence and stridency that seem to imply causation.

The title of Hamer's book is terribly misleading, as the book itself appears to accept: 'Just because spirituality is partly genetic', he says, 'doesn't mean it is hardwired.'[3] He also doesn't purport to say anything about *belief* in God – merely about some of the sensations sometimes associated with some people's self-proclaimed religious experiences. He doesn't deny that other biochemical pathways might be important in religious experience, or pretend that he understands the relationship between religious experience and belief. But these caveats haven't prevented him being savaged viciously by scientific opponents.

He didn't help himself by publishing his findings in a popular book rather than a peer-reviewed scientific journal, and subsequent peer review has been overwhelmingly negative. The gist was entertainingly summarised by Carl Zimmer, who said that the explanatory power of the VMAT2 thesis was so low that the book should have been called 'A Gene that Accounts for Less than One per Cent of the Variance Found in Scores on Psychological Questionnaires Designed to Measure a Factor Called Self-Transcendence, Which Can Signify Everything From Belonging to the Green Party to Believing in ESP, According to One Unpublished, Unreplicated Study'.[4] It is worth looking beyond the mud-slinging, though, to the small print of the objections. Some of them enlighten other areas into which we wander in the course of our journey through the biology of spiritual experience.

We must be very, very cautious about talking about genes 'for' various things. It is only very occasionally accurate to use that sort of language.

'It's nothing but modern molecular preformationism,' thunders P. Z. Myers, who is no friend of religion (it was Myers who caused international outrage by his desecration of the Eucharistic host).[5]

[It is] palmistry for the genome. We've been fighting against this simplistic notion of the whole of the organism prefigured in a plan or in toto in the embryo since Socrates, and it keeps coming back. We've moved from imagining a little homunculus lurking in the sperm to one hiding in the genome. It's just not there. You can't point to a spot on a chromosome and say, 'there's the little guy's finger!', nor can you point to a spot and say, 'there's his fondness for football!' [A supporter of Hamer] points to a particular gene as the source of piety. Piffle. Here's his shining locus of sacredness, VMAT2 . . . It's a pump. A teeny-tiny pump responsible for packaging a neurotransmitter for export during brain activity. Yes, it's important, and it may even be active and necessary during higher order processing, like religious thought. But one thing it isn't is a 'god gene'. The whole genome is like that. Browse VMAT2's neighbourhood, for instance, and you won't find 'philoprogenitiveness', 'benevolence' and 'chastity' marked out on the long arm of chromosome 10.[6]

What do you find? A long, complex list of proteins. 'It's like browsing through a Mouser electronics catalogue,' writes Myers, 'lots and lots of parts and pieces, but thin on the interesting bits about how you put them together to create a useful apparatus.' Where are the assembly instructions?

That information is all buried in the regulatory sequences that surround each gene, in the affinities for transcription factors for particular sequences of DNA, in the molecular interactions between proteins, in the pattern of environmental impact to embryo and adult, and in the whole unmapped history of the organism. VMAT2 isn't the answer; it's one among many parameters.[7]

You'll get things badly wrong if you don't realise that all organisms are far more than the arithmetical sum of their parts:

If small units (genes, neurones, whatever) added together in a linear fashion, then there'd be no problem breaking down complex organisms or behaviours into simple parts. But there are emergent properties at the organismal level that cannot be reduced in that way ... Phenotype isn't genotype. Instead, it's one-third genotype, one-third life history and (maybe) one-third something which, for want of a better term for it, might be called 'free-will', though I'm not introducing anybody's theology here.[8]

The phenotype/genotype distinction, basic though it is, is often disastrously submerged. We've all become so gene-focused that we can forget that the environment, and so natural selection, can't see genes at all. They see what genes *do*. 'Traits – including psychological and behavioural ones – are phenotypical, not genotypical,' Myers goes on, 'a marriage of countless different factors interacting with one another in countless ways. The genotype is a condensed and, to some extent, distorted map of the universe all the organism's ancestors lived in, and which the organism may have to deal with during its life.'[9] It is not the universe itself, as it is often taken to be. To confuse the map with the territory it purports to describe is a dangerously repercussive mistake. The VMAT2 story doesn't surprise Myers: 'There'll always be a market for easy explanations, even if they are wrong.'[10]

But all this means is that there is no one 'God gene'. It doesn't mean that one doesn't inherit a tendency towards being religious. Or non-religious. The twin studies, with all their theological corollaries, won't go away.

There's nothing particularly worrying about a tendency, though. We all have a greater or lesser tendency towards everything in the world, whether good or bad. Nobody's suggesting that Mick is somehow biologically incapable of belief. Indeed, all the evidence indicates not only that he is capable of it, but that he probably can't help having a set of beliefs in something. As indeed he has. He believes, for instance, with an intensity that would have been applauded in the Kremlin of the 1930s, that when he dies he will rot. That's a perfectly respectable belief, no different in kind from the belief that the Virgin Mary was herself conceived sexlessly, but supported by a great deal more evidence. He happens to have a lower tendency than some others, and than his local vicar would like, to apply specifically religious labels to his beliefs. I don't know how important or even how interesting that is. Some of that tendency nestles in his DNA – the product of an immensely complex interaction between lots of genes. And some nestles in his past – the product, very likely, of boring, insensitive proselytising by halitotic evangelists. But in the end, so what? Neurologically he seems free. Neither his genes nor his history predestine him either to heaven or to hell.

CHAPTER 4

Wholly Mad or Holy Madness?[1]

I had a religious delusion for about ten years . . . I was listening to music but . . . I was switching station a lot on the radio. Then I realised that the music titles were a hidden message especially for me . . . These titles ranged from 'Keep the Faith' by Bon Jovi, 'One Night in Heaven' by M People to 'God Gave Rock and Roll to You' by Kiss. There was a list of six songs that went to make up a message . . . I wrote to Bob Geldof and Bono telling them of my message and asking if they would arrange a concert for god . . . I was sectioned again . . . [T]he delusion became worse with hallucinations and voices as well. It was a very traumatic time for me. Then something in my perception changed, I was put back on the older Haldol [haloperidol] drug and the messages seemed silly, I turned to the church for under-standing and the delusion became my personal quest for faith. I converted to Catholicism and was confirmed. But nowadays I'm not so sure of my faith. I question the things I've seen as to whether they are real or simply a trick of the mind. . .

('Ebony', *Rethink*, 28 May 2008[2])

Gwen Tighe is a devout Christian and an epileptic. In 2003 she told the BBC *Horizons* programme how, during a seizure, she had given birth to Jesus. She experienced all the pangs of labour, saw him emerge from her and, when the seizure had stopped, was euphoric and flooded with spiritual enlightenment. It is not clear what she was enlightened about.

In about 1171 a twenty-one- or twenty-two-year-old Belgian peasant girl, Christina (later to be dubbed Christina the

Astonishing[3]), suffered a seizure so profound that she was thought to be dead. During her own funeral mass she came round and levitated to the ceiling of the church. The parish priest ordered her down. Being an obedient daughter of the Church, she obeyed, landed on the altar, and told the dumbstruck congregation that she had been to hell, purgatory and heaven. When her soul left her body, she said, it was ushered by angels to a very gloomy place, entirely filled with tormented souls. Like Dante, she recognised many of her acquaintances, and was moved by their agonies. She assumed that this was hell, but was told that it was purgatory. She was taken to hell next, where again she saw some old friends. Finally she went to heaven, where she tasted unspeakable raptures, saw the 'Throne of Divine Majesty', and was given a choice:

> Assuredly, my dear daughter, you will one day be with Me. Now, however, I allow you to choose either to remain with Me henceforth from this time, or to return again to earth to accomplish a mission of charity and suffering. In order to deliver from the flames of Purgatory those souls which have inspired you with so much compassion, you shall suffer for them upon earth: you shall endure great torments, without however dying from their effects. And not only will you relieve the departed, but the example which you will give to the living, and your continual suffering, will lead sinners to be converted and to expiate their crimes. After having ended this new life, you shall return here laden with merits.[4]

As in all the cases of near-death experience recorded in the literature, Christina opted to return. Until 1224, when she died of natural causes, she lived a life of relentless and relentlessly exotic misery, which she and her venerators thought was in some way redemptive. She slept on rocks, dived into furnaces and rolled in fires (apparently without harm), was savaged by dogs and sought refuge in thorn bushes (with no wounds), spent days standing in graves and icy rivers, and flung herself into a river and was whirled round by a mill wheel. She doesn't sound very likeable. She couldn't stand the smell of other people, because, she said, they stank of sin. To avoid the sinners she came to redeem she climbed trees, hid in cupboards or ovens, or levitated. In times of ecstatic trance she wandered shamanically between the worlds, leading the souls of the dead to purgatory, and souls who had served their purgatorial sentence to paradise.[5]

Christina the Astonishing, the twelfth-century levitating Belgian. Detail from a modern representation, by the US artist Cynthia Large, showing Christina levitated to a tree top.

The history of religion is crowded with epileptics.[6] Hippocrates called epilepsy 'the Sacred Disease', and in many cultures epileptics are revered as priests and spirit-walkers, able to penetrate the thin walls between the worlds. One could contend perfectly coherently that the history of religion was the history of temporal lobe epilepsy, or at least temporal lobe lability. Many key figures are known to have been, or may well have been, epileptic. Ellen

White, founder of the Seventh Day Adventists, had a brain injury at the age of nine, which is said to have transformed her personality. Epilepsy is a well-recognised sequel of traumatic brain injury, and it seems to have been a sequel in her case.

In her first visionary experience, in 1844, she saw the 'Advent People', dazzlingly illuminated, processing towards the New Jerusalem. She watched some fall from the path into Stygian gloom, the second coming of Jesus, and then the triumphal entry of the pure and steadfast into the Holy City. She reluctantly returned to earth to complete her mission. She was formally commissioned in another vision. 'While praying, the thick darkness that had enveloped me was scattered, a bright light, like a ball of fire, came towards me, and as it fell upon me, my strength was taken away. I seemed to be in the presence of Jesus and the angels. Again it was repeated, "Make known to others what I have revealed to you."'[7] Many other similar epiphanies followed, associated with bright white light.

Caravaggio's depiction of the conversion of St Paul. Was the Damascus road experience an epileptic seizure? (Cerasi Chapel, Rome)

Similar claims have been made for the pivotal experiences of even more pivotal people, such as Joan of Arc, the emperor Constantine (whose possibly epileptiform vision of the crucifix at the battle of Milvian Bridge converted him to Christianity, and caused the notional baptism of the subsequent Western world[8]) and St Paul. When we deal more generally with altered states of consciousness,[9] we will examine whether the visions of Ezekiel, John the writer of Revelation and other great biblical figures might have been biologically comprehensible. If you are searching for the inspiring spirit of some of the Bible, temporal lobe epilepsy has to be on the list.

And not just of the Bible. Whether or not epilepsy is the well-spring of their creativity, a striking number of the world's great artists have been epileptic. They include Dostoyevsky, Edgar Allen Poe, Gustav Flaubert, Lewis Carroll and Sylvia Plath – all people, when you come to think of it, with the sort of icon-oclastic *élan* that might come from having looked back at this world from the perspective of another.[10]

Some epileptics, indeed, have made an explicit connection between their condition and religious experience. Dostoyevsky, giving his own experience in the words of Prince Myshkin, the epileptic character in *The Idiot*, spoke for them all:

I have really touched God. He came into me myself: yes, God exists, I cried. You all, healthy people can't imagine the happiness which we epileptics feel during the second before our attack.

A patient studied by Saver and Rabin reported that 'my mind, my whole being, was pervaded by a sense of delight'. Another told of feelings of detachment, ineffable contentment and fulfil-ment. He visualised a bright light, which he recognised as being the source of all knowledge. Sometimes, too, he saw a bearded young man, whom he thought was Jesus Christ.[11]

Fyodor Dostoyevsky: 'I have really touched God,'
he wrote, describing his own experience of epilepsy.
(Vassilij Grigorovi Perov, 1872)

This sort of story is very common, says the neuroscientist Vilayandur Ramachandran. He claims that 25 per cent of temporal lobe epileptics will have some sort of religious epiphany (often described as 'seeing God' or 'gaining enlightenment') just as they are being swept off into a seizure. But sufferers (or should they be 'the blessed'?) are not just hyper-religious when they are in or approaching a seizure. Even when they are 'normal', they have a greater than average emotional response to religious words and symbols.[12]

The maverick neurobiologist Michael Persinger, who pursues his largely self-funded work in Laurentian University, Ontario, and who wears a three-piece suit and fob watch even when he mows the lawn, agrees that electrical activity in the temporal lobes is the explanation for religious experience. He is explicit about the mechanism.[13] He thinks that unusual impulses in the right temporal lobe compel the rather anoraky left hemisphere to concoct an explanation. The explanation that the left comes up with is understandable but wrong: a real presence which, depending on the subject's cultural and religious background, might be labelled God, Jesus, the spirit of a departed relative, the Great Spaghetti Monster, or inchoate awe.

He has famously sought to demonstrate this by his 'God helmet'. This has passed electromagnetic fields of varying patterns through the temporal lobes of many hundreds of subjects. The psychologist and paranormal researcher Susan Blackmore, subjected to patterns similar to those seen in temporal lobe epileptics having religious experiences, wrote that her initial scepticism soon dissolved. First she felt herself swaying, as if she were on a hammock.

> Then it felt for all the world as though two hands had grabbed my shoulders and were bodily yanking me upright. I knew I was still lying in the reclining chair, but someone, or something, was pulling me up. Something seemed to get hold of my leg and pull it, distort it, and drag it up the wall. I felt as though I had been stretched half way up to the ceiling. Then came the emotions. Totally out of the blue, but intensely and vividly, I felt suddenly angry – not just mildly cross, but that sort of determinedly clear-minded anger out of which you act – only there was nothing and no one to act on. After perhaps ten seconds it was gone, but was later replaced by an equally sudden fit of fear. I was just suddenly terrified – of nothing in particular. Never in my life have I had such powerful sensations coupled with the total lack of anything to blame them on.[14]

Another woman thought that her dead mother was standing by her side. The British journalist Ian Cotton 'realised' that he was and had always been a Tibetan monk.[15]

But not everyone is impressed. Richard Dawkins's leg twitched and he felt slightly dizzy, but he did not meet his judge. The English writer John Horgan 'waited for God or even a minor deity or demon to appear – in vain'.[16] Persinger shrugged off their disappointment, on the grounds that the helmet doesn't seem to work with real sceptics – a riposte that seems tantamount to an acknowledgement that suggestion plays at least some part in the electronic epiphanies.[17] But not necessarily. Dawkins had low scores in psychological tests that predicted epiphanies: perhaps the scepticism picked up by those tests was itself an index of temporal lobe stability. Persinger claims that about 80 per cent of subjects 'sense a presence' when they have the helmet on, compared to 15 per cent in a control group.[18] Most of these experiences, though, were far from spectacular. Charles Cook, a former graduate student of Persinger who was involved in God helmet procedures in the 1990s, has said that most of the 'positives' had only a vague sense of being watched. They were indeed being watched like hawks – by the researchers.[19] Swedish researchers purported to replicate Persinger's experiments under double-blind conditions: they could not verify his conclusions.[20] The rhinoceros-skinned Persinger dismissed their dismissal, saying that the Swedes did not replicate his experiments at all: they failed to expose the subjects to the necessary fields for long enough.

The jury is still out on the God helmet. There is depressing prejudice in the scientific community against what is widely seen as off-the-wall research. The prejudice has denied Persinger any significant funding: it may deny him a fair hearing. That would be a tremendous shame. Whatever one thinks about his methods and his conclusions so far, the work is of colossal importance. And there are tentative reasons for supposing, because of what we know entirely independently about temporal lobe instability, that he might, just might, be on to something.

'Sensed presence' wavelengths aren't the only ones that can pulse through the God helmet. Persinger is developing a whole electromagnetic pharmacopoeia. If he's right, a sequence called 'Burst X' can induce a pleasant, relaxed sensation. Another one, called the 'Linda Genetic Pulse' after one of his assistants, is said to boost the body's natural self-healing powers.[21] He claims to have found a pattern that, on rats, has the analgesic effect of morphine, and he hopes to be able to diminish depression and control epileptic seizures. He thinks that his work has ousted the traditional God of Sinai, but thinks that he might be able to fill the gap and install a bespoke-tailored God in our living rooms. 'Can we use [the helmet technology] to decrease the anxiety in an increasingly secular world?' he asks.

> People dying of cancer, who don't believe in God – we could use that stimulation to allow the feeling of wholeness, to allow the feeling of personal development. In the future, you may find a space in the average home, much like in the Eastern tradition, which is basically your God centre, where you sit down, 'expose' yourself – it may not be a helmet by then – where you would be able to pursue your own personal development. Do we have a technology here that will allow us to pursue the last greatest mystery, which is your own introspection?[22]

One of his associates, the neuroscientist Todd Murphy, thinks so, and thinks that the market will agree. Murphy is already marketing a version of the helmet, which he calls the 'Shakti machine', as a sort of cognitive cross-country vehicle, for 'consciousness exploration'.[23]

Temporal lobe epilepsy is one of many diseases that can induce experiences labelled by their experiencers as religious. Some have been touched on already: any condition that stimulates the amygdala, for instance, can theoretically trigger awe or fear.[24] Some might say that since life itself (a terminal condition with almost a 100 per cent mortality rate[25]) carries with it the creeping terror of death and the consequent creeping or explosive desire to transcend death, life itself is a religious-experience-inducer. Death and its attendants (critical ischaemia, cardio-respiratory failure and a flat EEG) certainly can be. A chapter of this book is devoted to experiences on the very threshold of eternity.[26] Privations such as pain, exhaustion, dehydration, sensory deprivation and extreme cold can usher subjects into altered states of consciousness in which beatific things can happen. They too have their own chapter.[27] But there are other medical chariots of the gods: they include brain bleeds, migraine, schizophrenia and 'Jerusalem syndrome'.

A little while ago I flew to Chennai and climbed stickily on the train south. We rumbled through paddy fields and waving forests of bananas and, from Tiruchirapalli, lurched on in a badly lame Ambassador cab to an ashram by the banks of the sacred River Cauvery – the Ganges of southern India. Here, in the hot dark, fanned by flickering bats, and in the steaming light, looking up at the swifts diving among the palms, I chanted the *Gayatri* mantra in Sanskrit – the ancient priestly language of India.

> *Om bhur bhura svata*
> *Tat savitur varenyam*
> *Bhargo devasaya dhimahi*
> *Dhigo yo nah prachchodyat*

Salutations to the Word which is present in the earth, the sky and that which is beyond. Let us meditate on the glorious splendour of that divine Giver of Life. May he illuminate our meditation.

And then we moved on to the Eucharist, for this was a Christian ashram. It is called Shantivanam, and is famous as the home of Dom Bede Griffiths, an English Benedictine monk, pupil of and fecund correspondent with C. S. Lewis. Here he tried to arrange a marriage between East and West; between intuition and reason; between the right brain and the left brain. If you're a conservative Protestant, you'll think that he tried to arrange a marriage between light and darkness, truth and falsehood, and God and Satan, by using the time-honoured technique of complete capitulation.

In the afternoons, when it was too hot to read, pray or sleep, I used to go to the tiny hut where Bede lived and died. There is nothing there now apart from his bed (he castigated himself for never being able to sleep all the time on the floor) and the table where, in patient, scholarly, spidery prose, he tried to articulate his holistic vision of Christianity as *advaita* – non-duality.

Lying on that bed, not long before he died, he suffered a stroke. He felt pressure mounting in the left side of his head, propelling his brain (as he put it) towards the right – towards the feminine, intuitive side. It was the greatest moment of his life. It enabled him to 'Surrender to the Mother' – to the force of unconditional, accepting love which he, as a constitution-ally tweedy, inhibited Englishman, had always desired and from which he had always run. The Mother crashed in and swept him off. He cried out to a friend nearby, 'I'm being over-whelmed by love.' And so it seemed. Even the cynics talked about his transfiguration. He described it as the triumph of the feminine in him, and was amused that his sexuality had taken so long to kindle. And so a celibate monk was prepared for death.

*Bede Griffiths, the English Benedictine monk and friend of
C. S. Lewis, who spent most of his life in India. Not long before
he died he suffered a stroke. He thought that this propelled his
brain to the right – towards the feminine, intuitive side – and
saw it as the culmination of his life's searching.*

Neurologically this story is interesting. It is conceivable that
a scan would have shown exactly what Bede thought had
happened: that a bleed in the left, logical, linguistic hemisphere
handed executive jurisdiction to the holistic, intuitive right hemisphere in a way that a lifetime of strenuous spiritual exercises
had failed to do.

The German abbess Hildegard of Bingen has perhaps done
more than anyone to build bridges between the Middle Ages
and the contemporary world. Her music is always near the top
of classical bestseller lists; her esoteric visions feature in many
a New Age anthology of feminine mysticism. We meet her
again in a context in which she would be outraged to appear –

the chapter on holy eroticism – but let's look for a moment at what a modern neurologist would make of her.

From a very early age she was often ill.

> God punished me for a time by laying me on a bed of sickness so that the blood was dried in my veins, the moisture in my flesh, and the marrow in my bones, as though the spirit were about to depart from my body. In this affliction I lay thirty days while my body burned as with a fever, and it was thought that this sickness was laid upon me for a punishment. And my spirit also was willing, and yet was pinned to my flesh, so that while I did not die, yet did I not altogether live. And throughout those days I watched a procession of angels innumerable who fought with Michael and against the Dragon and won the victory . . . And one of them called out to me 'Eagle, Eagle,[28] why sleepest thou? . . . All the angels are watching thee . . . Arise I for it is dawn, and eat and drink' . . . And then the whole troop cried out with a mighty voice . . . 'Is not the time for passing come? Arise, maiden, arise!' Instantly my body and my senses came back into the world; and seeing this, my daughters who were weeping around me lifted me from the ground and placed me on my bed, and thus I began to get back my strength. But the affliction laid upon me did not fully cease; yet was my spirit daily strengthened . . . I was yet weak of flesh, timid of mind, and fearful of pain . . . but in my soul I said 'Lord, Lord, all that Thou puttest upon me I know to be good . . . for have I not earned these things from my youth up?' Yet was I assured He would not permit my soul to be thus tortured in the future life . . . Three years were thus passed during which the Cherubim thus pursued me with a flaming sword . . . and at length my spirit revived within me and my body was restored again as to its veins and marrows, and thus I was healed.[29]

But only in her 'veins and marrows', not in her head. And even in her veins and marrows, not for long. Throughout her life she was often prostrate. Her illnesses were the gateway to strange visionary worlds. She was not asleep during the visions, she insisted, nor mad.

> These visions which I saw I beheld neither in sleep, nor in dream, nor in madness, nor with my carnal eyes, nor with the ears of the flesh, nor in hidden places; but wakeful, alert, with the eyes of the spirit and with the inward ears, I perceived them in open view and according to the will of God.[30]

Flames from on high engulf the head of Hildegard of Bingen, the twelfth-century German visionary. Her ecstatic visions sound like manifestations of scintillating scotoma.

The visions were very various. Sometimes they sounded as if they came straight out of Revelation, Ezekiel, Daniel or an ayahuasca séance with an Amazonian shaman.[31]

I looked and behold, a head of marvelous form . . . of the colour of flame and red as fire, and it had a terrible human face gazing northward in great wrath. From the neck downward I could see no further form, for the body was altogether concealed . . . but the head itself I saw, like the bare form of a human head. Nor was it hairy like a man, nor indeed after the manner of a woman, but it was more like to a man than a woman, and very awful to look upon. It had three wings of marvelous length and breadth, white as a dazzling cloud. They were not raised erect but spread apart one from the other, and the head rose slightly above them . . . and at times they would beat terribly and again would be still. No word uttered the head, but remained altogether still, yet now and again beating with its extended wings.[32]

But despite their variety, there are many recurrent features. They all have a central point or group of points of light. This central focus moves, shimmering or undulating, and is often interpreted by Hildegard as burning eyes or stars. Often the largest light has concentric, rippling rings of light around it. Her lights are generally active: they seethe or ferment.[33]

'How this was compassed', wrote Hildegard, 'is hard indeed for human flesh to search out.'[34]

Well, possibly. But Singer, with whom the neurologist Oliver Sacks and many others agree, thinks that at one level the answer is easy enough: 'the medical reader or the sufferer from migraine will, we think, easily recognize the symptoms of "scintillating scotoma".'[35]

Is that the end of the story? Does a diagnosis answer all questions about the origin and significance of the visions? The answer to both questions is a resounding 'no'. We return to the questions later. But there is no point in pretending that the diagnostic label does not make neurological sense, nor in denying that it can sensibly be applied to many other experiences upon which whole theologies have been built.

We all have visions of a sort. My son Tom said, 'When I press my head hard onto my cot, I see a DVD.' He was describing the geometric patterns, sometimes known as entoptic phenomena, which are by-products of the natural structure of the eye. If you press on your eyeballs you may see flashing lights, dots, flecks, sets of parallel lines or zigzags. The phenomena are common, too, when coming out of sleep, and in the antechamber of many altered states of consciousness. Deeper inside those altered states the entoptics might seem to take on a life of their own: to swirl and writhe. We will meet them again and again in this book. They are anatomically interesting; they are psychologically dull. Unlike schizophrenia.

An artist's depiction of entoptic phenomena – geometric patterns generated by the natural anatomy of the eye. In altered states of consciousness they can assume special significance. (Aldo)

Today, schizophrenics are caged and their ectopic voices pharmacologically stifled. But in cultures more ancient and broad-minded than our own, the voices heard by a schizophrenic may be regarded as the whisperings of the gods.

All humans were schizoid once, according to Julian Jaynes,

who articulated the thesis of the 'bicameral mind' – the idea that until about three thousand years ago the human mind was functionally divided into two.[36,37] The right hemisphere was the director. The left was an executive. The right hemisphere ordered the left hemisphere around, communicating with it through auditory hallucinations. When the ancients wrote about hearing the voices of the gods, this wasn't just a figure of speech: they really heard directive voices which they attributed to the Olympians, the Muses, the Logos, Yahweh or their tribal gods. There was no consciousness of self during these times, says Jaynes. Just look at the total lack of introspection in Homer or the older parts of the Old Testament. 'Homer does not "do" inner life (not much, anyway – we do get the odd flashes from Odysseus' deep soul),' writes Charlotte Higgins. 'By and large, characters are sketched out through what they say and do, not what they think and feel. His characters rage when they are angry, weep when they are unhappy, and speak their minds. There are no resonating psychological layers; Penelope is not Isabel Archer.'[38] Jaynes noted, too, that the ancient world was crawling with gods, and they were much more anthropomorphic than the few that replaced them. That's just what you'd expect, he said: the gods were no more than the voices of the right side. There were as many Mount Olympuses as there were right hemispheres.

Jaynes thought that during the second millennium BC, as societies grew more complex, the bicameral mind began to break down. Societal complexity was fuelled by and itself fuelled the increasing ability of humans to generate and experiment mentally with different models of the world.[39] The right-hemispherical gods were bludgeoned almost to extinction by the symbol-wielding warriors of the left. Many would say that language was in the vanguard of the vanquishing army. During the breakdown phase,[40] divination and prayer arose in an effort to extract instructions from the newly silent gods. People who retained bicameral function would be consulted as seers, prophets and oracles. At first, children continued to be able to commune with the gods, but they quickly evolved their way into the adult isolation from 'them'.

There are vestiges of bicamerality today. Not all bicameral individuals languish in secure psychiatric units. Auditory hallucinations, even in the apparently sane (by which is meant, with some irony, an ability to put on a suit and sit at a desk), are surprisingly common.[41]

Jaynes's ideas have received some support from neuro-imaging studies. The conventional neurological wisdom is that, so far as language is concerned, the right hemisphere is fairly inert, but when auditory hallucinations occur, activity can be seen in the parts of the right hemisphere that correspond to the language centres in the left.[42,43] But there are some very potent objections. The first and most fundamental is that it is a theory built almost entirely on the absence of an expressed interest in human consciousness from a very small number of old documents of uncertain literary form. If you don't talk all the time about the problem of consciousness, it doesn't make you unconscious. It just makes you a regular bloke. And detectable introspection needn't intrude into every conversation, nor into every piece of writing. We know little about the literary and bardic conventions that structured Homer and Hesiod, and until it is established that they would have made their characters introspective if they had been capable of it, it is rather harsh to diagnose bicamerality.[44]

There are also grave chronological difficulties with Jaynes's ideas. The history of the Upper Palaeolithic is, classically and overwhelmingly, the history of increasingly sophisticated symbolic thought. Symbols coat the walls and fill the graves of the Upper Palaeolithic. Symbols need not necessarily connote consciousness, but they suggest it, and it is a bulwark of Jaynes's own theory that bicamerality cannot survive long in the presence of symbolic thought. *The Epic of Gilgamesh* – the original version of which is well before Jaynes's watershed second millennium – is an embarrassment to him. It seethes with introspection. He shrugs and says simply that the introspection was added later. It is too convenient. One might as well contend that all the introspection was edited out of the *Odyssey* on stylistic grounds. Similar embarrassments face him whenever an anthropologist wanders into a village relatively untouched by modernity. There are plenty of people in Amazonia, New Guinea and elsewhere who live lives hugely less sociologically complex than the lives depicted in the *Iliad*. And their right hemispheres don't talk audibly to their left.

If vestigial bicamerality isn't a driver for modern religious oddness, what is?

Perhaps places can be.

An American man in his forties, with a history of paranoid schizophrenia, started working out frenetically. As his deltoids swelled, he started to think of himself as the biblical character Samson. He became obsessed with the idea that one of the gigantic stone blocks in Jerusalem's Western Wall was in the wrong place, and so he got on a plane to Israel, got the bus to Jerusalem, and tried to move the block. There was a predictable confrontation with religious Jews, and he found himself in Jerusalem's Kfar Shaul mental hospital – an institution to which a surprising number of religious tourists are carted off each year.[45] An unwise psychiatrist challenged him, pointing out that the Bible recorded no visits to Jerusalem by Samson, and that for various other (probably even more compelling) reasons he could not possibly be Samson. The patient was furious, smashed a window and escaped through it. He was eventually coaxed back to the hospital, where he was found to be acutely psychotic. He responded well to anti-psychotic medication and was able to fly back to the US.[46]

A South American Protestant plotted to destroy Islamic holy places so that they could be replaced by Jewish buildings. He then planned to start the war of Gog and Magog, forcing the Antichrist to reveal himself and so hastening the second coming of Christ. He gutted a mosque in Jerusalem. Psychiatric examination showed that he was unable to distinguish between right and wrong, and that he was accordingly unfit to stand trial. He was incarcerated in an Israeli psychiatric hospital before being sent home.[47]

A forty-five-year-old German man, with no previous history of psychiatric illness, became obsessed with a search for the 'true' religion. For five years he marinated himself in the sacred texts of Christianity and the religions of ancient Persia, China and Japan, before concluding that none of them was the 'one'. He left his academic job and went to Jerusalem, where he studied Judaism at a university and a yeshiva, before rejecting that too. He eventually decided that the only true religion was 'primitive Christianity' – the elusive creed preached by Jesus and corrupted by Peter and Paul. He saw himself as a divinely appointed ambassador of this revelation, and preached it wherever and whenever he could, in heavily accented English. In the Holy Sepulchre Church in Jerusalem, the supposed site of the death, burial and resurrection of Jesus, he became violently agitated, shouting at the priests and accusing them of worshipping graven images and being pagans and barbarians. They didn't take it well. A physical struggle erupted, and before the self-appointed prophet was carried off to Kfar Shaul he managed to damage some statues and paintings. Subsequent psychiatric examination showed no trace of any psychiatric illness. One might have expected at least a personality disorder, but none was found. He just had a fixed idea that he had understood the ancient writings properly – a fixed idea shared by a depressingly high proportion of the world's religious enthusiasts. Three years after this, he was back in his old job in Germany, his convictions unchanged, sad only that he could not preach to the spiritually blind and unregenerate in the Holy City.[48]

For each of these patients Jerusalem was a stage on which to act out their perceived parts in the drama of redemption. Perhaps we all have such stages: some perform in churches, temples and mosques, others in hospitals, soup kitchens, families, pubs, books or the labyrinthine cloisters of their own heads. But for another class of patients, Jerusalem itself seems to be a trigger for some fascinating weirdnesses.[49,50]

It normally goes like this. A Western Protestant with a deeply uninteresting previous medical history arrives in Jerusalem as part of a tour group. On his first night he seems anxious, agitated and nervous. Then he breaks away from the group, becomes obsessive about bodily cleanness (repeatedly showering and cutting his fingernails and toenails), makes himself long white robes from his hotel bedsheets, and wanders around Jerusalem (particularly the Old City), shouting, screaming or singing religious lyrics or verses from the Bible. He then marches ceremonially to one of the holy places (often the Church of the Holy Sepulchre or the Western Wall), where he delivers a loud, confused and anguished sermon, exhorting his listeners to turn from their wickedness.

About a week after the start of the episode he is probably on the way back to normal, whether or not he has had the benefit of the copious experience of the Kfar Shaul psychiatrists. Afterwards he can recall everything that he has done, and is deeply embarrassed by it. He describes himself as behaving like a 'clown' or a 'drug addict', but is reluctant to be pressed hard about what happened. If he does talk he might describe a sense of 'something opening up inside him', creating an obligation to do or proclaim something.[51]

The patients who behave like this are almost always Protestants (40 out of 42 in a study by the Israeli psychiatrist Bar-El) from

ultra-religious families for whom the Bible was the most important book, and who read the Bible together at least once a week.[52]
The Bible was the root of their lives; its pictures were their most
treasured images. They defined themselves by reference to the
Bible. Perhaps, then, when the cherished images were contradicted by the real Jerusalem, they suffered some sort of self-
eroding angst that had to be dealt with by a dramatic reassertion
of the old, picturesque Jerusalem of their childhood Bible, in
which everyone wore togas and sandals, and prophets bellowed
at every street corner. Bar-El comments:

> [To sufferers from this type of Jerusalem syndrome] [t]he Bible
> would ... serve as a source of answers to seemingly insoluble
> problems ... For fundamentalist believers of this type, Jerusalem
> assumes the highest significance: such people possess an ideal-
> istic subconscious image of Jerusalem, the holy places and the
> life and death of Jesus. It seems, however, that those who succumb
> to [this type of Jerusalem syndrome] are unable to deal with the
> concrete reality of Jerusalem today – a gap appears between their
> subconscious idealistic image of Jerusalem and the city as it
> appears in reality. One might view their psychotic state and, in
> particular, the need to preach their universal message as an attempt
> to bridge the gap between these two representations of Jerusalem.[53]

We have seen before how the left hemisphere clings desperately
to the models of the world it has created. It hates to have their
inadequacy demonstrated. It hates even more the sight of them
being slung onto the tip. It resents the voice of the right hemisphere ordering it to build something that fits better with the
observed world. It loves its self-drafted pastiches of the truth
much more than it loves the truth. The right hemisphere is more
honest. I suggest that in Jerusalem syndrome we see a classic
battle of the hemispheres. The left hemisphere has constructed,
with the help of the King James Version, a view of the universe

with which it is inordinately satisfied. The right hemisphere looks at the real Jerusalem and says, 'Sorry, it's not like that. Back to the drawing board.' The left hemisphere throws its toys out of the pram and puts on a show in a pathetic attempt to convince itself and the right hemisphere that the old view was correct all along. Often it wins: the fundamentalists go back to the States and to their well-thumbed Bibles with their prejudices defensively reinforced and more immune than ever to future attacks from the real world. Indeed, the greater the evidence against the position, the greater the conviction with which the position is held. This is seen in many cults[54] and exemplified by the apparently increasing strength of the Creationist movement.[55] Contradiction by the available evidence is seen as an attack by a demonic enemy. Hasn't God told them to expect attack? So strong attacks are seen as strong divine confirmation that they are following the Right Path. Psychology has given the phenomenon a name: Cognitive Dissonance. If my suspicion about the battle of the hemispheres is correct, that is spot on. It is a dissonance between the left hemisphere and its models on the one hand, and the right hemisphere and the real world on the other.

If religion is about finding the truth, the Jerusalem ravings aren't religious experiences at all: they are anti-religious experiences. Quite a lot of notionally religious experiences are like that.

I don't at all dismiss the idea that there is something objectively holy about certain places, and I have tasted the neurological potency of holiness. When you stand at a Jerusalem bus stop you feel as Lucy in *The Lion, the Witch and the Wardrobe* must have felt as she rushed up to the wardrobe for the second time. You feel as if there's only a very thin film that separates you from another world – and it's not just because a brainwashed teenager might board the bus with a knapsack full of Semtex. I am perfectly happy to believe that there is something in Stonehenge, the old ley lines of England, Lourdes, the sacred

71

mountain of Arunachala, Walsingham, the Ganges or Ayres Rock that resonates ecstatically with the most fundamental fibres that run though human beings. Is there something in feng shui? I don't know, but it wouldn't surprise me a bit. There are some places with no obvious associations in which I can work productively and be happy, and others in which I fidget, write nonsense and struggle exhaustingly against the resident gloom. It would be strange if it were otherwise – if our psyches were hermetically sealed from our environments. Our relationship with this planet is unimaginably, vertiginously old, and unimaginably intimate. We are built with its elements, and I share the overwhelming majority of my defining DNA with the other organisms on it. Even the bacteria in my gut are relations. Perhaps supposedly 'holy places' communicate their 'holiness' to me through the medium of an electromagnetic field. Does the demonstration of the medium tell us anything about the sender or the message? Well, not much. It certainly doesn't tell us that there is no sender. We return to this thought in the final chapter of this book.

So much for illnesses that can cause religious experiences. What about religious experiences that have physical corollaries?

It is perhaps worth mentioning generally that if there is any correlation at all between being religious and being healthy, it seems to be a positive correlation.[56] At the time this book was written advertisements were appearing on London buses saying, 'There's probably no God. Now stop worrying and enjoy your life.' Many of the sponsors of the ad were scientists who should have known better. There is no evidence that religious thoughts torment more than they comfort. When it suits them, the atheistic sponsors are perfectly happy to explain away the existence of religion by saying that it confers a survival advantage by stopping people worrying distractedly about their own deaths. They can't have it both ways.

But if religion itself is good for you, the effect is very slight. I argue in Chapter 11 that any effect is far too subtle to be picked up and championed by natural selection. The studies that tend to show an advantage in being religious are beloved of those Christian apologists who don't see the Darwinian downside of relying on them, but they are difficult to interpret. People who are religious will, by and large, have a lower incidence of smoking, drinking and risky sexual behaviour than the non-religious. Because religion tends to be a sociable activity, they are also more likely than non-religious people to have an emotionally and physically supportive network of friends. Strip out these factors and most of the perceived health benefits of being religious will evaporate.

But that's being religious, not having religious experiences. It's methodologically tricky to distinguish between them, but some religious experiences do seem to be good for you. A study of 1,000 British evangelical clergy concluded that the 80 per cent who spoke in tongues were more emotionally stable and less neurotic than the 20 per cent who did not.[57] Other studies have not replicated those findings, but certainly *glossolalia* has never been shown to have any detrimental effects on mental or physical health. Sociologist Andrew Greeley, using standard psychological benchmarks, found that people who claimed to have experienced mystical states had a 'state of psychological well-being substantially higher than the [US] national average'.[58]

This is mild stuff. There are two obvious, direct connections between religious experience and physical health. First, madness generated by religious experience, and second, stigmata.

First: madness. The literature doesn't contain any dramatic examples of people seeing things that they perceive as religious and as a result being catapulted into psychosis. While Richard Dawkins would no doubt say that any religious faith generated by any religious experience is necessarily an example of precisely that, he'd be silly to do so. The connection between religious experience and religious faith is curious, complex and equivocal: we examine it later.[59] There are plenty of examples of already psychotic people channelling their psychosis into religion: some cases of Jerusalem syndrome, for instance. And of course schizophrenics will often assert that they are a major religious figure in whatever culture they have been brought up. A Catholic schizophrenic may become convinced that he is Jesus. A Hindu schizophrenic may think he is Ganesh. But that doesn't mean that the religion has caused the schizophrenia. No doubt many Muscovite schizophrenics in the 1930s devoutly believed that they were Stalin. People are driven mad by hunger, pain and loneliness, but rarely by a vision of the Virgin. Debilitating depression – the sort that makes you mumble through the street and drink a bottle of vodka in the morning – is much more likely to be caused by a feeling of the absence of God than by a feeling of his presence.

Second, and finally, stigmata.

'I bear on my body the marks of Jesus,' wrote St Paul.[60] While we have no real idea what he meant by this, many (mostly Roman Catholics) have claimed the same, and Christian stigmatics almost always mean that they have developed physical signs or symptoms in sites corresponding to the wounds received by Jesus at or around his crucifixion. Some show wounds to the hands or wrists, the feet or the side; some have shoulder wounds, as if from carrying the cross; others have scourge marks to the back, or scalp wounds as if from the crown of thorns; others claim to sweat or weep blood. Sometimes the blood (often said to be a mixture of the subject's blood and Christ's) will not clot; sometimes the wounds heal very soon after they first appear. If medical treatment is sought (as it sometimes has been), it generally does no good, but if the wounds do not heal, as sometimes they do not, they are said never to get infected. Indeed, the blood oozing from them may perfuse the room with a holy perfume – the Odour of Sanctity. Sometimes there are no visible wounds, but the subject will feel pain in the prescribed places.

The hands and feet of St Francis are pierced by an angel. (Giotto, c. 1325, Bardi Chapel, Santa Croce, Florence)

Stigmata generally appear after an ecstatic, often visionary, religious experience. St Francis' stigmatisation was fairly typically dramatic. One morning in 1224, when praying with even more fervour than usual, he saw a six-winged angel. The angel came closer, and he saw that it had been crucified. A wave of joy and agony burst over him, and when it retreated he discovered wounds in his hands, feet and side.

The famous Italian stigmatic Padre Pio wrote:

It all happened in a flash . . . I saw before me a mysterious Person . . . His hands, feet and side were dripping blood. The sight of Him frightened me: what I felt at that moment is indescribable. I thought

I would die, and would have died if the Lord hadn't intervened and strengthened my heart which was about to burst out of my chest. The Person disappeared and I became aware that my hands, feet and side were pierced and were dripping with blood.[61]

The wounds were obvious, and Padre Pio was embarrassed by them:

I am dying of pain because of the wound and because of the resulting embarrassment which I feel deep within my soul . . . Will Jesus who is so good grant me this grace? Will He at least relieve me of the embarrassment which these outward signs cause me?[62]

But there was to be no relief. He bore the wounds for fifty years, eventually, apparently, coming to cherish them. They bled profusely and continuously. If it was a grace, it was a strange and sometimes resented grace.

The Italian stigmatic Padre Pio. For fifty years his hands, feet and sides oozed blood.

Stigmata are often associated with other mystical phenomena. In some of the stories there is the sense that they are badges of honour – medals given to reward faithful service.

In Damascus, in November 1982, an eighteen-year-old Christian girl, Mirena, was praying for a very sick friend, along with two other women. One of the women noticed that light was streaming and oil flowing from Mirena's hands. She shouted at Mirena, telling her to put her hands on the sick woman. Mirena did, and the woman was immediately and completely healed. It was the start of a strange mystical career.[63]

A few days later a small picture of the Virgin and Child in a cheap plastic frame in Mirena's house began to ooze oil, and oil again began to seep from Mirena's hands. Baffled relatives met to pray with Mirena, but as they prayed aloud, Mirena could hear nothing. She got up and put her ear to the dresser on which the picture stood. 'Do not be frightened,' she heard. 'I am with you. Open the doors and do not stop anyone from seeing me . . . Light a candle for me.' And so she did. The effects were dramatic. Not only did oil continue to pour out of the picture and from Mirena's hands, but also out of other pictures of the Virgin brought to Mirena's house, or blessed by her. Mirena and the pictures effected spectacular healing miracles, and the Virgin appeared to Mirena several times.

It was about a year later, in October 1983, that Mirena began to feel nails in her hands. Oil came out of her face, neck, chest and hands, her body became rigid and cold, and at the beginning of November red spots appeared on her palms. They progressed to open wounds, and foot wounds followed. Mary comforted her: 'Do not be afraid; all this happened so that the name of God will be glorified . . . [T]hrough your example I will educate this generation.' Some of Mirena's Marian ecstasies (in which she sees a curiously cold, jealous

Mary) seem to have some of the classic hallmarks of out-of-body experiences.[64]

> I found myself in the clouds, and I saw my Mother Virgin Mary. She was smiling at me . . . [Her] smile changed and became gloomy. She said: 'Go down and tell [your family and friends] that you are my daughter before being their daughter. . .' I repeated [to my family and friends] what she said. I saw them all crying around me. Also I saw my body lying down on the bed.[65]

Stigmata are not uniquely Christian phenomena. If you walk long amongst the tutelary spirits of the Waraw people of the Orinoco delta, for instance, the spirits may, for theologically obscure reasons, bore holes in your palms.[66] Caitanya Mahaprabhu, the sixteenth-century Bengali mystic who popularised the chanting of the Hare Krishna mantra, to be heard on many busy Western shopping streets, is said to have oozed blood during times of particularly intense yogic contemplation. But some of the non-Christian 'stigmata' loosely referred to in the literature are not stigmata as the Christians understand them at all. 'Buddhist stigmata', for instance – the thirty-two signs of a Buddha – are bodily characteristics (mostly indices of perfect proportion) that are supposed to mark out an Enlightened One. They indicate a supposed concordance between physical and spiritual perfection. Christian stigmata, with a deafening paradox that some will find exhilarating and others incomprehensible, point up a concordance between spiritual perfection and running wounds.

Cynicism about stigmata is easy, common and often well founded. But there is a danger of throwing out the genuinely mystical baby with the psychosomatic bathwater. Yes, the first recorded stigmatics (unless you include St Paul) appear in the thirteenth century, when the medieval Western world first became obsessed with the physical wounds of the crucified Jesus. As the images proliferated, so did the stigmatics. Yes, female stigmatics have historically outnumbered males by a ratio of about seven to one (perhaps a reflection of the lowly status of women in the Catholic Church: having a colourful clutch of stigmata was one of the very few ways a woman could be noticed). Over the last century, during which women have become less systematically suppressed, the ratio has changed: it is now about five to four in favour of women. Yes, stigmata are generally a Catholic phenomenon, and Catholics do have a particular interest, if not preoccupation, with the wounds of Christ. In Catholic churches the gaping mouths of the statues' anatomically accurate wounds are meant to speak eloquently to the worshipper. In Protestant churches, for better or for worse, the gibbet is empty. Yes, the sites of the stigmata do seem to depend on the stigmatic's understanding of where Jesus' wounds were. Wide publication of the suggestion that the Turin Shroud showed nails through the wrists rather than the more traditional palms led to a minor epidemic of wrist wounds amongst devout Catholics. Yes, stigmatics are often people with extraordinarily (one might say pathologically) acute imaginations. The stigmatic Heather Woods, who took down in uncannily fast handwriting sermons dictated to her by God, and had very regular visions of Jesus (especially being baptised and crucified), said that she stood beside Jesus and John the Baptist in the Jordan River: 'I could see the water dripping from [Jesus], sparkling in the sunlight.'[67] Yes, as we have seen, stigmata are often associated with dramatically altered states of consciousness. Yes, there is a well-documented phenomenon of hysterical conversion. A very mild form is described in Jerome K. Jerome's *Three Men in a Boat*, when the narrator goes to the library and starts to read a medical encyclopaedia. He concludes, to his horror, that he has every disease in the book except housemaid's knee. 'I walked into that reading room a happy healthy man. I crawled out a decrepit wreck.'[68]

But it can be a lot more serious than that. Wounds, not just feelings of wounds, can be generated by intense identification with the wounded.[69, 70] It is not surprising that compassion for and identification with the crucified Christ can leave imprints not just on the psyche but also on the body.

But why should the label 'hysterical conversion' or any of its allies in the psychiatry textbooks drain the phenomenon of stigmata of its awe? Of course, as evidence of any of the axioms of the Christian faith, stigmata are hopeless. No one should look at Mirena's feet and say, 'They convince me of the truth of the resurrection,' or, 'Now *there's* a good reason to believe in God.' But nor should anyone look at her feet and conclude that all mystical phenomena have a hysterical aetiology. The uncontroversial conclusion would be that the corporeal and the incorporeal are potently and intimately related. And if you can agree with that most bland of propositions, you're halfway to being a mystic.

CHAPTER 5

Getting Out of Yourself:
An Introduction to Other
States of Consciousness

What is man? asks young Werther – man, the glorious half-god?
Do not his powers fail him just where he needs them most? Whether
he soars upwards in joy or sinks down in anguish, is he not always
brought back to bald, cold consciousness precisely at the point
where he seeks to lose himself in the fullness of the infinite?

Often I have thought of the day when I gazed for the first
time at the sea. The sea is vast, the sea is wide, my eyes roved
far and wide and longed to be free. But there was the horizon.
Why a horizon, when I wanted the infinite from life?

(Thomas Mann, *Disillusionment*, trans.
H. T. Lowe-Porter, 1896)

It was five o'clock on a Sunday at a big charismatic church in
London's exclusive Knightsbridge area. A two-bedroomed flat
in the parish would cost you over a million pounds. Just down
the road Harrods glittered and strutted. In smaller shops round
the corner you could buy a bit of the Parthenon, a Turner sketch
or a medieval Turkmen carpet. In the restaurants manicured
couples ate goose livers and patisserie flown in from Paris. The
baubles round the necks of the customers in one coffee shop
would have discharged the national debt of Laos or paid for a
small but perfectly formed coup in Togo.

The church was packed to its Victorian rafters. Most of
the people were young. They were a mix of professionals,

students, reformed and unreformed addicts, City workers, rehabilitated bank robbers who had met God in prison, and curious tourists. There was a buzz of meeting, gossip and excited expectation.

The band came onto the stage, plugged in, and the congregation stood up. At the first cadences people shut their eyes, raised their hands and swayed.

'There must be more than this,' sang the leader. 'O breath of God, come breathe within. . .'

More than what? Why did they think that life has anything more to offer than what they had? It was plain from the look on their faces that they weren't here just to be taught doctrine. They were here to meet something or someone. Some of the raised hands were like the hands of a supplicant beggar. Some punched the air triumphantly to celebrate or endorse a point in the song. And some were stretched up and out in longing, as if to embrace a lover on a railway platform. Sometimes a hand would come down to wipe away a tear. Sometimes lips would move in a language known by no dictionary.

Up in the balcony a group of amused bystanders pointed and sniggered – but rather nervously, as if they were not quite sure of the grounds for their derision.

Science has generally sniggered too. The American psychologist William James wrote:

No part of the unclassified residuum [of human experience] has usually been treated with a more contemptuous scientific disregard than the mass of phenomena generally called mystical. Physiology will have nothing to do with them. Orthodox psychology turns its back on them. Medicine sweeps them out; or, at most, when in an anecdotal vein, records a few of them as 'effects of the imagination' – a phrase of mere dismissal, whose meaning, in this connection, it is impossible to make precise. All the while, however, the phenomena are there, lying broadcast over the surface of history.[1]

Not only over the surface of big, broad, human history, but also over the surface of our own personal histories. The worshippers in Knightsbridge wanted something else. Harassed commuters in overcrowded trains read books or listen to music that takes them to somewhere else. Gyms thrive because everybody wants to live in a body other than their own, or experience the serotonin rush that comes after strenuous exercise. In the gorgeous Knightsbridge houses, immaculate through the labours of Filipino house-slaves, corks pop as bottles of ancient Burgundy are opened to open the gates to other states of consciousness. And if that doesn't work, there's the perpetual, doomed hope that Cuban cigars, Columbian cocaine or initially exciting adultery might. Mozart plays on the music centre; it might shift the EEGs from the beta waves needed in the office to generate the income to pay the mortgage, to more mellow alpha or even theta waves. The children, before they are packed off to Eton and Wycombe Abbey, are raised on stories of other worlds – worlds accessed through wardrobes, picture frames, holes in the ground, caves, space rockets, cyclones or magic words. None of us seems entirely at home here. We all suspect that there is more than this. Not only do we want to know what else there is; we want to taste it. We want to go there.

The least radical suggestion about what else there might be is that there might be a portfolio of other types of consciousness. Aldous Huxley wrote:

> The ordinary waking consciousness is a very useful and, on most occasions, an indispensable state of mind, but it is by no means the only form of consciousness, nor in all circumstances the best. Insofar as he transcends his ordinary self and his ordinary

mode of awareness, the mystic is able to enlarge his vision, to look more deeply into the unfathomable mystery of existence.[2]

We love, idolise and loathe ourselves, thought Huxley. Much of our loathing stems from boredom. We look at ourselves and say: 'Now he is really, truly, emetically dull. And yet he is the god I worship. My idol makes me sick.' Hence the desire for self-transcendence – for escape from the insufferable tedium of our own company. We desperately try various escape routes: yoga, meditation, life-threatening mountains, piety, drugs, good works, bad novels, great novels, sex and death.

Huxley mistakes the real nature of transcendence. He has listened to too many bellowing public school chaplains asserting that St Paul wants us all to crush our real selves to death, and has mixed their mistake with a dash of Buddhism to make a heady but deadly cocktail. But that's what they wash down with their Burgundy in those Knightsbridge houses.

Nobody seriously doubts that there are other types of consciousness. You can buy one for the price of a bottle of whisky, a bag of cannabis or a wrap of ketamine. But there are gentler ways of getting one.

'I'd love to have a go at hypnotising you,' said Sally, at a dinner party. Sally is a doctor with an almost clairvoyant diagnostic talent, a hypnotherapist, and a very good friend. There aren't many people I would trust to rummage through my insanitary psyche, but she is one of them. And so, a couple of weeks later, filled with misgivings, cynicism and downright fear, I cycled up to her house in north Oxford. I was scared of disappointing her by being unhypnotisable. I was scared of being too hypnotisable, and giving away embarrassing things. Although I'm not at all impressed with where I have steered my own mind, I was scared of relinquishing control. I was scared of flinging open gates that are normally kept shut for a reason, and being invaded by horrors. I was scared of seeing the writhings inside my head. I was scared of too much clarity.

'I'd like you to think of a safe place,' Sally had told me. 'A place where you have been completely happy. I'd like you to visualise every detail of it. Think too how it sounds and smells. I needn't know where it is, but you need to be able to retreat into it.' Retreat from what? I wondered. And has there ever been a place of complete safety and complete happiness?

I lay down on the sofa, shut my eyes, and tried to relax. Sally told me to think nice things, which was easier said than done, and to take deep breaths. Very slowly my muscles thawed out and my mind began to stop hopping between projects and neuroses. Sally was talking all the time: gently, kindly and reassuringly, but without obvious suggestion. Twice she asked me to open my eyes and follow her finger as it moved up and down my field of vision. 'Now,' she said, 'I want you to take five breaths, as deep as you can.' I did. 'Now I want you to start at 100, and count backwards, very slowly, to 96. Perhaps take one number per breath. Perhaps one number every two breaths. As you say each number, think of it as fireworks bursting in the sky. Each burst is slightly smaller than the one before. By the time you get to 96 there is almost nothing to see.'

I hit 96, and nothing obvious happened.

'Now think of yourself in a summer garden. You are walking down a path. You can feel the gravel under your feet and the sun on your face. You can smell the roses and hear the birds. Stop at a flower bed if you like; see the shape of the petals and watch the bees.' I was there.

'There is a door from the garden. It goes to your safe place. Walk through it.'

I went to where I thought I was safe. But I wasn't. Sally saw it before I did, but her voice didn't change. She called me quietly back into the garden. Downstairs I could hear her husband coming in from walking the dog. I knew that if I opened my eyes I would see the computer, and the ranks of DVDs, and Sally sitting by the sofa, watching for the flickering eyelids that indicate a trance-like state. 'Find somewhere else. There are many such places. Think of holidays with Mary and the boys, perhaps.' And then something odd, and oddly familiar. 'You are being filled from the toes upwards with peace. It goes to the top of your head and beyond.'

Behind my closed lids my eyes were wobbling like jelly flicked hard by unseen fingers. I found a refuge, but soon the knowledge that it was there was enough: I didn't need to stay there. I liked it in the garden.

Sally wanted me to go back to an old, unhealed wound. 'Think of your life as a line,' she had said before we started. 'Does it stretch straight out in front of you, and behind you, or what?' I told her. 'And are you standing directly on it?' I wasn't, and told her where I was. She seemed interested. We went back along that imagined line, sometimes queasily fast, sometimes too slowly for comfort. Sometimes I looked away from a sight along the line, hoping, red-faced, that Sally hadn't seen it too. The sore bit was a little weal: it doesn't really bother me. But Sally held my hand metaphorically and we picked over it together; rationalising things; seeing things without the distortion of perspective that comes from twenty years; hovering directly over it; and then, at last, fast-forwarding and fast-reversing several times over it, as if to erase it, before accelerating away back to the present. I went back to the garden. In the kitchen the dog barked and a machine whirred: Tony was making a cake. I didn't want to leave the garden. There were lots of interesting things to see, and I wanted a picnic on the grass bank. But the hour was up. 'Three, two, one: and open your eyes slowly.' Tears poured out. I wondered how my eyelids had held them all in. I was more fully in my body than

I ever remember being before. From the reaction downstairs I must have looked stunned and dazed. I felt neither. I felt as if I had slept for forty-eight hours.

As I cycled away down the Banbury road, I asked myself what my predominant state of mind was, expecting an answer like 'peaceful', 'relaxed', 'focused', or even 'puzzled'. I was surprised and mystified by the answer that came: I felt an overweening *pity*. Pity, that is, in the sense of seeing suffering that I hadn't seen on my cycle up, empathising with it and wanting, desperately and hopelessly, to do something about it. I didn't even feel nauseated by such sanctimonious, hypocritical thoughts in myself, as I usually would. I went there expecting something interesting and probably vaguely good for me; I found something *good*.

That's not to say that it wasn't interesting. It was. Nobody really knows what is going on in hypnosis, although there is no lack of theorising. It is certainly not unconsciousness. Instead the consciousness seems to be concentrated inside one's own mind, or in the self-generated summer garden. Peripheral awareness is correspondingly reduced, although the dog was loud and undistorted. The state resembles sleep 'only superficially', and is 'marked by a functioning of the individual at a level of awareness other than the ordinary conscious state. This state is characterized by a degree of increased receptiveness and responsiveness in which inner experiential perceptions are given as much significance as is generally given only to external reality.'[3]

Functional MRI studies reveal that highly susceptible subjects show, under hypnosis, greater activity than weakly susceptible subjects in the anterior cingulate gyrus (which evaluates emotional outcomes and responds to errors) and the left side of the prefrontal cortex (involved in higher-level processing and behaviour).[4] These findings are perhaps rather counter-intuitive.

They suggest that in hypnosis we are more critical than normal; that we are functioning both emotionally and intellectually at a higher level than usual; that we are better integrated than when we're sitting at our desks. In short, that we are *more* ourselves when in a hypnotic trance, not less. And where are we to find our real selves? Where is hypnosis taking us? 'Hypnosis is a phenomenon of degrees, ranging from light to profound trance states but with no fixed constancy,' we are authoritatively told. 'Ordinarily, however, all trance behaviour is characterized by a simplicity, a directness, and a literalness of understanding, action, and emotional response that are suggestive of childhood.'[5] Again we meet the idea of the child as the *real*; the child as the real inheritor of whatever kingdom is out there.

At the end of the service at the Knightsbridge church, anyone who wants to can come up to the front to be prayed for by a member of the prayer ministry team. It's man on man, woman on woman, to avoid any nasty misunderstandings and to reduce embarrassment. The pray-er asks the pray-ee what they'd like prayer for, puts their hand on the pray-ee's shoulder, and prays. The prayer will normally involve an invitation to the Holy Spirit to come. Indeed, the whole exercise is underpinned by the theological understanding that whatever happens, happens through the power of the Holy Spirit. Sometimes things happen physically to the pray-ee during the prayer. They may move in an agitated way; they may sometimes fall to the ground (the phenomenon known as being 'slain in the Spirit', in which subjects describe being wrapped blissfully in the love of God, enjoying him, and being unaware of how time is passing). But more commonly their eyelids will flicker, as mine did on the couch, and tears will come. Boxes of tissues are put out in anticipation. The tears and the eye movements are taken as manifestations of the Holy Spirit, and who am I to say that they are not?

Afterwards subjects often report that their burdens have been lifted; that love has flooded them; that they have peace. Physical healings are common – but then hypnosis is reliably reported to improve physical well-being too. But to observe a similarity is not to establish a connection, let alone identity. We can't help but notice, though, that trance states are common in the New Testament, and that important visions are often given to subjects while in a trance.[6]

What happened on the north Oxford sofa, at the front of the church, and possibly in the worshipping congregation too, seems to have been the induction of another state of consciousness. If our usual state of consciousness is to be equated with this world, I, the pray-ees and the worshippers went to another one. We seem to be separated from those other worlds by the thinnest of veils – nothing like as robust as a wardrobe door. The day after my hypnosis I managed to entrance myself on a Bakerloo Line train between Paddington and Embankment.

For most of human history, though, there have been specialists in travel between worlds: the shamans. Their work presumes that there is a spirit world intimately interdigitating with and affecting our own. The spirits affect us profoundly. They bring rain, plague, death, glut and military triumph. The shaman can shuttle between this world and others, wrestling with or appeasing spirits, releasing the souls of animals to the plains of our world to improve the hunting, retrieving the exiled soul of a diseased person or curing infertility by snatching the soul of a child so that it can be implanted in a woman's uterus. They may have agents or fixers in the spirit world whom they meet on their journeys, and who help them to get what they want.

Some shamans are born; they will travel the routes their fathers and forefathers trod. Others are made by the epiphanic effect of disease or a near-death experience, or anointed by a lightning strike. The voices in a schizophrenic's head may be heard as a commission from the spirits. Where substances such as ayahuasca or peyote are the primary vehicles for travelling between worlds, easy accessibility to the other realms can diminish the status of the shaman, but still there will be an ayahuasca-master or peyote-priest, a veteran of many voyages, who will titrate the drug to the individual subject, may act as a guiding Virgil in the far country, and can debrief the returning travellers, expounding what they have seen.

An Alaskan shaman exorcising an evil spirit from a boy, 1890s. (Library of Congress, Frank and Frances Carpenter Collection)

The shamanic universe is laminated. There is a world below and a world above this one, but the relationship between them is neither simply spatial nor straightforwardly non-spatial. The boundaries between the worlds are porous, but you need to know where to look for the portals. The portals may have some real earthly correlates – hence the shamanic interest in caves, holes and rivers. Many places in Greece and Turkey were thought to be the forecourts of Hades. Goblins come from the centre of the world and blink in the light of Middle Earth. When the Inuit shaman goes to the underworld he falls through a spiralling, plughole-like vortex in the sea. The deepening trance of a Kalahari shaman is described as a voyage underwater and through the depths of the earth.[7] Aslan comes from across the sea. But we should be careful not to patronise, ascribing a naivety to shamanic cultures that they did not have. Subtle truths can get lost as we caricature. A good example is the ascension of Jesus. The story is told in Acts.

> So when they had come together, they asked him, 'Lord, is this the time when you will restore the kingdom to Israel?' He replied, 'It is not for you to know the times or periods that the Father has set by his own authority. But you will receive power when the Holy Spirit has come upon you; and you will be my witnesses in Jerusalem, in all Judea and Samaria, and to the ends of the earth.' When he had said this, as they were watching, he was lifted up, and a cloud took him out of their sight. While he was going and they were gazing up towards heaven, suddenly two men in white robes stood by them. They said, 'Men of Galilee, why do you stand looking up towards heaven? This Jesus, who has been taken up from you into heaven, will come in the same way as you saw him go into heaven.'[8]

Laughable, say the sceptics, pointing to those ridiculous paint-ings of amazed apostles looking upwards, baffled, as the feet of Jesus disappear into a cloud. And the sceptics invite us to conclude from this that the ascension couldn't have happened and that the book of Acts is a crude fairy story. But this is classic straw-manning. Did Jesus' feet really peep out of a cloud? I doubt it very much. The Jews of the time weren't so dim as to think that heaven, or wherever it was Jesus went to, was physically 'up there', where the rain came from. 'Up there' was a conventional way of describing somewhere thoroughly real, but different from the place that we're in at the moment. As a metaphor it is not bad, but, like all metaphors, has its shortcomings. It probably tells us more (and certainly told first-century Jews more) than any more metaphysically sophisticated attempt at an illustration would have done. Perhaps other worlds are more accurately described in terms of parallel universes or warps in Einsteinian space–time, but those descriptions are no use to me. Anyway, it is a shame that the Bible is too often read and illustrated by dyed-in-the-wool metaphor-phobes.

The literature of religion, myth and downright entertainment (but why are we entertained by it?) is full of journeys between universes. The ancient world was awash with dying and rising gods and quasi-gods: Balder, Dionysus, Persephone, Osiris, Adonis, Jesus and so on. Heroes penetrated and plundered under-worlds (for instance Orpheus, Theseus and Jesus). Genies could be summoned by esoteric incantations, or the rubbing of lamps. Flying carpets crowded the skies of Arabia and Persia, competing with the winged discs of Sumer, chariots of fire of the sort that abducted Elijah, the strange ship with its mysterious crew in the book of Ezekiel, various fairy vehicles, witches on broom-sticks and, later, flying saucers.[9] Medieval Europe undulated with levitating saints.[10] Sci-fi is the spice of life for millions, and pays the mortgages of thousands of publishers, writers and bookshop owners. As a boy at a very English public school, I was taken on thought-journeys by a bearded, orange-clad master who had served his apprenticeship at the Poona ashram of the notorious Bhagwan Shree Rajneesh. The sixth-form general studies class called 'Who am I?' sat in an oak-panelled room with their tweed jackets off and their grey-serge-trousered legs crossed. We left the playing fields, the river and the Oxbridge honours boards behind, and soared to a crystalline lake in a blue tropical forest. We lay on the back of a hippo, ate teacakes, wrestled with a bear, escaped by turning ourselves into coconuts, and were back in time for double chemistry.

Modern shamanic ceremony, Siberia.

Since then I have travelled to many places by staring into candles; by trying to hurl out the thoughts as they bombard me like hailstones; by chanting old prayers in the dark. Sometimes I have been taken. Sometimes I have watched my own mind as it flutters from flower to flower, and thought that each flower was a world. And sometimes, but very seldom, I have been in all the worlds at once.

Modern shamanic ceremony, Siberia.

There are rougher, cruder ways to travel. Perhaps they are more reliable, but perhaps the journey is indistinguishable from the destination. Some psychoactive drugs will suck you down that vortex; some exercises involving pain, exhaustion or the elimination of sensation will catapult you through the veil. Others seem to prise 'you' out of your body without killing you. Profound cerebral anoxia can lift you out of yourself so that you hover over your hospital bed looking down interestedly, but usually peacefully, at the crash team working on your chest.

We look in detail in other chapters at the means of travel and the landscapes on the other side of the veil. But a word of warning. If you read the mystical literature, particularly of the East, you will see that the pearl beyond price is the state beyond being: Absolute Unitary Being – the abolition of all distinction between you and the rest of the universe; the death of dualism; the state of complete oneness with everything else. To get there, the Upanishads and other teachings say, you need to know,

Modern shamanic mask, Siberia.

really *know*, that the only real thing is the Void. It is infinite, formless and without personality. It spawned us and everything else. We return to it. But here's the joke that the Zen Buddhists have got, and which makes them smile. In fact, since we are of the Void, and it is *not*, we do not exist at all. There *is* no ground of our being, and we have no being. So be not afraid: you cannot die. To experience the reality of this fundamental Unreality is to be free of hope, and therefore free of fear. The Holy Grail of eastern mysticism is not only empty, it is Emptiness. And it's not holy.

We have very few reliable reports from people at one with the Void. Perhaps we have none. But few have reached whatever there is to reach. The knowing smile on many Zen faces is not a smile of real knowing. But this is not the impression that you get from the books in the alternative bookshops. It is easy to get the impression that Absolute Unitary Being is just a matter of going along to ta'i chi every Wednesday in the community hall. It's not. And there are many casualties along the way. Perhaps, if Absolute Unitary Being really is the true Grail, that's what you'd expect. But the *type* of casualty should give pause for thought. There are real horrors in the look over the edge. If you transcend everything, you may find that you've transcended benevolence.

There's another important and related misrepresentation in the literature. It is that the real, gilt-edged mystical experiences are all strenuously, dogmatically, emphatically, by definition impersonal and amoral. But that's not how the dispatches read. Far more commonly we read of encounters with a *presence*. And a presence that is *good*. Those accounts are perhaps under-represented in the literature because they sound too cringe-makingly *religious*. An experience of *bhakti* is cool, and therefore discussable in a university common room. But an experience of being met and hugged by a Person: now that's too toe-curlingly revivalist to warrant anything but a patronising sneer.

The Jewish High Priest offering a goat as a burnt sacrifice. Many religions have priestly intermediaries standing between the deity and the people. The priests often make metaphorical journeys into the presence of the deity on behalf of the people, akin to shamanic journeys. (Henry Davenport Northrop, Treasures of the Bible, *1894)*

CHAPTER 6

Finding God in a Garden:
How Psychoactive Substances can
Throw Open the Doors of Perception

One is nearer God's heart in a garden
Than anywhere else on earth.

(Dorothy Frances Gurney)

Badger, badger, badger, badger, badger, badger, badger, badger,
Badger, badger, badger, badger, *Mushroom*, *Mushroom*,
Badger, Badger, Badger ... *Snake*, *Aargh*, *Snake* ... Badger,
Badger, Badger.

(Flash movie animation)[1]

I have seen UFOs split the sky like a sheet ... I have seen seven
balls of light come off a UFO, lead me onto their ship, explain
to me telepathically that we are all one and there is no such
thing as death ... I lay in a field of green grass for four hours
going: 'My God ... I love everything.' The heavens parted, God
looked down and rained gifts of forgiveness onto my being,
healing me at every level, psychically, physically, emotionally. And
I realised our true nature is spirit, not body, that we are eternal
beings.

(Bill Hicks, of his experiences on Psilocybin mushrooms)[2]

The stink of the petrochemical plants of modern Elefsina hits
you as soon as you drop down from the crest that separates
Elefsina from Athens. On the Sacred Way, along which the devo-

tees of the Eleusinian Mysteries processed, there are crane-hire companies, mechanics' workshops, motor showrooms and sex shops. It's a dual carriageway for much of the way. The last bit of the Sacred Way in Eleusis is now a scruffy street running past a coach park. Perhaps appropriately, it's a one-way street.

The site of ancient Eleusis is a huge acropolis, almost dwarfed by the industrial chimneys rising up at one end of the escarpment. They look like monstrous Doric columns. At the chimney end of the mound you can take a rough track that climbs steeply up, through olives and cypresses, to the tiny chapel of St Nicholas, squashed between the perimeter of the archaeological site and the domes of another chemical works. It's as near to the sacred site of Demeter's sanctuary as you can build. The message is obvious: these are the real Mysteries, and they've supplanted the old.

The sanctuary of Demeter, at Eleusis, near Athens.
Here thousands in the ancient world lost their fear of death.

At the foot of the hill, in a quiet garden, a girl was reading a letter and crying. There were piles of empty glue pots on the hill. The adolescents of Elefsina obviously carry on the ancient tradition of artificial induction of altered states of consciousness.

Eleusis was where the ancient world lost its fear of death. 'Happy is he', wrote Pindar, 'who, having seen these rites, goes below the hollow earth, for he knows the end of life and its god-sent beginning.'[3] For some it gave not just confidence, but life beyond the grave: 'Thrice happy are those mortals who, having seen these rites, depart from Hades,' thought Sophocles, 'for to them alone is granted to have a true life there. For the rest, all there is evil.'[4]

When the Roman Empire became Christianised, the Eleusinian Mysteries were outlawed. But for over a thousand years the priests of Demeter celebrated her successful mission to the underworld to snatch back her daughter Persephone. The temple of Demeter had no windows. It must have been rather like a coffin, and perhaps deliberately. As the devotees filed into the temple they were given a drink called *kykeon* made, we're told, from water, barley and mint. We do not know what they saw there. To divulge the secret meant death. But perhaps Persephone appeared as a column of light, disgorged by the great dark, embodying the rebirth of the earth in spring, a dead seed springing into a flower so dazzling that all the terrors in the gloomy souls of men fled screaming before it.

When the initiates came out of the temple after the rite, they were quiet and reflective. They came out alone and silent, or they spoke gently in small groups. No doubt many of them went down to stare out at the sea. The road to the harbour is now full of pretentious restaurants, clubs and a disproportionate number of hairdressers, all of them with a lot of chrome. But the sea is the same.

Persephone, helped by Hermes, emerges from the underworld into the arms of her mother, Demeter. (Sir Frederick Leighton, 1891, Leeds City Art Gallery)

About two thousand years and eight hundred miles separate Eleusis from Basel, in Switzerland. In Basel, in 1943, the chemist Albert Hofmann felt agitated and dizzy. He had been working on a substance like ergot – a fungus that can contaminate grain. Wondering if the substance might be responsible, he took a very small dose in water. His world fell apart. Chairs became threateningly malevolent; neighbours became masked witches. But the evil dissolved, and the world, newly created, shimmered with love and light. Lucy was in the Sky with Diamonds, and all was for the best in this best of all possible worlds. He had discovered LSD.

It has been suggested that the grain used in the *kykeon* was contaminated with ergot, and that the initiates of the Eleusinian

Mysteries were on an acid trip. We don't know: the debate rages.[5] But it is at least plausible. Far less plausible is Dan Merkur's notion that the wilderness wanderings of the Israelites were cheerier than the book of Exodus paints them because the manna was laced with ergot.[6]

The use of psychedelic drugs is immeasurably old. A cave painting in the Tassili mountains of Algeria, dated at 8000 BC, shows a strange figure, identified as a mushroom-using shaman.[7] The cave paintings of Upper Palaeolithic Europe, going back around 35,000 years, may contain coded evidence of hallucinogens,[8] mushroom motifs and statues in Mayan temples in Guatemala may indicate a ritual use of magic mushrooms, and perhaps the journeys of the ancient Egyptian dead, presided over by divine animal–human hybrids, were journeys done and illustrated by Egyptian shamans who used the blue lotus to shuttle between worlds.

Every continent has seen psychoactive drugs. The traditional peoples of Australasia may not have used psychoactive drugs, but any trip to the clubs of Sydney demonstrates that modern Australians feel the need for them. And I'll bet there's been the odd E dropped or spliff smoked on some of those grim Antarctic bases. Wherever hallucinogens have been used, they have been seen as a door to a world worth investigating. Many have seen them as the gate of Wisdom or of heaven itself. For many, St Peter, the guardian of the heavenly gate, is a drug dealer.

In Gabon, Cameroon and the Democratic Republic of the Congo, a preparation of the bark of the iboga tree brings devotees of the Bwiti religion face to face with God. 'To see God,' proclaims the Bwiti priest in the liturgy, 'we must eat the body of God symbolized by iboga, and the cithara, Ngoma, takes us by the hand and leads us towards God . . . it . . . takes us from the here and now to the beyond, from the profane world to the sacred world, from the world of the living to that of the dead.'[9]

The drums pulse, the iboga is taken, the worshipper lies on the floor looking up at the roof of the temple or the fruit bats flapping across the forest canopy. The iboga spirit often materialises as a ball of dazzling light. 'Do you know who I am?' he typically asks. 'I am the chief of the world, I am the essential point!'[10] 'I had numerous visions and ponderous metaphysical insights,' wrote an English journalist who had become a Bwiti initiate. 'At one point I seemed to fly through the solar system and into the sun, where winged beings were spinning around the core at a tremendous rate. Up close they looked like the gold-tinged angels in early Renaissance paintings.'[11] You can see why people wonder about Ezekiel and St John of Patmos.

The peyote cactus has apparently been used sacramentally in North and Central America for at least 5,500 years, and is still used by the Native American Church (who have negotiated an exemption from the usual draconian US laws governing the use of psychoactive drugs).

A typical service in the Native American Church will last about ten hours. Hymns will be sung in one of the old languages, lots of tobacco in corn husks will be smoked, someone will thump a deerskin drum, lumps of peyote will be eaten and an infusion of peyote will be passed around and drunk from a communal bowl, for all the world like an ordinary Christian Communion service. Some people will vomit discreetly; others will weep as the peyote disinhibits them. The doses taken are generally small – not enough to produce the spectacular visions so often written about. John Halpern, a Harvard psychiatrist who has been working for years on the sacramental use of peyote, thinks that peyote 'serves primarily as an amplifier of emotions aroused by the ceremony's religious and communal elements'.[12]

Much bigger doses of mescaline (the psychoactive ingredient in the peyote cactus) famously changed the life of Aldous Huxley (who always had a nasty tendency to over-philosophise).

Mescaline shook him out of the despised ruts of ordinary, workaday perception, and showed him, 'for a few timeless hours, the outer and inner world, not as they appear to an animal obsessed with survival or a human being obsessed with words and notions, but as they are apprehended, directly and uncon-ditionally by Mind at Large. . .'.

'The experience is neither agreeable nor disagreeable,' he wrote. 'It just is.' He was seeing

> [t]he Being of Platonic philosophy – except that Plato seems to have made the enormous, the grotesque mistake of separating Being from becoming and identifying it with the mathematical abstraction of the Idea. He could never, poor fellow, have seen a bunch of flowers shining with their own inner light and all but quivering under the pressure of the significance with which they were charged; could never have perceived that which rose and iris and carnation so intensely signified was nothing more, and nothing less, than what they were – a transience that was yet eternal life, a perpetual perishing that was at the same time pure Being, a bundle of minute, unique particulars in which, by some unspeakable and yet self-evident paradox, was to be seen the divine source of all existence.[13]

Mushrooms have certainly been important in Central American religion for many centuries, even if one doesn't accept the evidence for their use in the Mayan civilisation. A small model of a mushroom looking very like the psychoactive *Psilocybe mexicana* has been found in a shaft and chamber tomb in western Mexico. It has been dated at AD 200, and its funerary context strongly suggests some ritual use. There is no doubt at all that the Aztecs used hallucinogenic mushrooms in religious rituals. They called them 'God's mushroom' or 'flesh of the gods' – a description which the devout, brutal and apparently easily shocked Conquistadors took to be disgustingly blasphemous.

Whenever the Spaniards felt the need to justify their bestial suppression of native religion, they trotted out the description as if nothing else needed to be said.

Indigenous South Americans have a huge psychoactive pharmacopoeia, of which the most famous example is ayahuasca – 'the Vine of the Dead' – used still in forest huts in Amazonia (sometimes with a congregation of curious Western psychedelic tourists, usually earnest graduates from the marijuana high school, anxious for a slice of enlightenment) and by middle-class, suburban Brazilian members of the Uniao de Vegetal, a religious cult that uses ayahuasca as its sacrament. It is always used together with another Amazonian plant which acts as a monoamine oxidase inhibitor. If the inhibitor isn't used, the active ingredient in the ayahuasca, dimethyltryptamine (DMT) is broken down too quickly to have much effect. It has rightly been asked how on earth the shamans knew which one of the many thousands of plants in the Amazonian rain forest would have the monoamine oxidase inhibitory effect. The shamans' own answer is that the knowledge didn't come from earth at all: plant spirits told them. Neither science nor anthropology can yet supply a more satisfactory answer.

A jungle séance always takes place at night, supervised by a shaman who is an expert in the geography of the land to which ayahuasca takes the subject, and who determines the dose according to the subject's experience. In the hut, leaning against a wooden post, listening to the shrieks of the night and surrounded by the shaman's votive objects, the English journalist Graham Hancock sipped ayahuasca from a chipped cup.

Slowly, after some false psychic alarms, he found himself wandering through an anteroom into another verdant world. He was surrounded by intelligent plants – so intelligent that they were almost like animals. He stroked a huge boa constrictor as if it were a house cat, marvelling at the reticulated pattern on

its back, and murmuring, 'You're a beauty.' An enormous butterfly
led him to another great snake which radiated 'sentience and a
magical force'. As he looked, the snake turned into a jaguar. But
perhaps most oddly he met small humanoids, made of white
light. They had heart-shaped faces, their eyes were completely
black, with no pupils, they seemed to have some telepathic business with him, and he felt that he was possessed by the spirit
of his dead father – about whom he felt a crushing guilt.

A subsequent session crawled with snakes, which he felt were
intelligent and benevolent. But it was not all good. He saw the
alien-type faces again. They seemed malevolent, and he feared
abduction. A dragon-snake changed into a colossal insect-like
creature with humanoid features. He saw a vision of the earth
rotating gently in space between two cupped hands. He wondered
at a ravishingly lovely Egyptian goddess and a male human
figure with the head of a crocodile. And what, he rightly asked,
would they have made of all these experiences in the Upper
Palaeolithic?[14,15]

In India, soma, a hallucinogen of unknown identity, had a
crucial mind-opening function, according to the Vedas. 'We
have drunk the Soma; we have become immortal,' sings the *Rig
Veda*. 'We have gone to the light; we have found the gods. What
can hatred and malice do to us now ... The glorious drops
that I have drunk set me free in wide space ... Weaknesses and
diseases have gone; the forces of darkness have fled in terror.
Soma has climbed up in us, expanding.'[16] 'One of my wings is
in the sky; I have trailed the other below. Have I not drunk
Soma! I am huge, huge! Flying to the cloud!'[17] 'You speak of
the sacred ... you speak of the truth ... Where the inextin-
guishable light shines, the world where the sun was placed, in
that immortal, unfading world, O Purifier, place me ... in the
triple dome, in the third heaven of heaven, where the worlds
are made of light, there make me immortal.'[18]

Soma may have been an extract of the fly agaric mushroom, *Amanita muscaria*,[19] a potently psychoactive fungus certainly central to Siberian shamanism and probably ubiquitous in Asian and European shamanism too. The *Rig Veda* says that the potency of soma increases in the urine (how did anyone find *that* out for the first time?), and that is certainly true of the fly agaric mushroom. When the Siberian shaman takes it, he saves his urine and others drink it, with splendid hallucinogenic effect. That really is what getting pissed is all about. The effect lasts until the compound has passed through five sets of kidneys, and then diminishes.

The fly agaric mushroom may have had profound effects on modern Western culture,[20] but it may also have had one less profound but still indelible one. Have you ever wondered why Santa Claus wears red and white, has reindeer, flies through the air and comes in and out through a chimney? It's because he's a mushroom. The fly agaric has a distinctive red and white canopy, and when taken by the Siberian shamans it propelled them on a shamanic journey out through the smoke hole in the roof of the yurt to fly with their spirit animals – reindeer – to the realms beyond. The shaman returned to his couch by coming back through the smoke hole, bearing gifts from the spirits.[21]

Amanita was probably well known to European shamans. Robert Graves and others[22] have contended, credibly, that ambrosia, the immortality-conferring food or drink of the Olympian gods, was a psilocybin mushroom such as *Amanita*. Some, noting that ambrosia and nectar are often used almost interchangeably, have thought that they are both references to fermented honey – mead – and that the celebration of ambrosia in Greek myth records the start of the Dionysiac alcohol cult. The gods guarded ambrosia jealously – hence the tragic story of Tantalus, punished for the crime of giving ambrosia to the mortals by being immersed to his neck in water, only to have it drain away when he bent his head to drink.[23]

Apollo may have had his own chemical vehicle to relay his messages to men. His oracle at the mountain sanctuary of Delphi, squatting on the sacred tripod and pronouncing judgements on which the ancient world turned, may have been inhaling ethylene, methane, carbon dioxide or hydrogen sulphide, spewing from deep in the earth through a volcanic fault.[24,25] Plutarch noted that the temple was filled with a sweet smell,[26] which could have been ethylene, a reasonably potent trance-inducer.

Whatever the identity of ambrosia, the classical world was certainly not short of hallucinogens. The lotus-eaters, nine days storm-blow from the tip of the Peloponnese, lured two of Odysseus's crew into a chemical stupor so delightful that they did not want to leave the island, and had to be dragged back to the ship and tied down to go cold turkey.[27] Lotus may have been the jujube fruit, the mandrake (apparently important in Egyptian religious ceremonial), or one of many other candidates.

Through a plastic curtain at the back of a small café in a side street of the Andriyivskyy Descent in Kiev is a covered courtyard. There, for a modest fee in US dollars and an exorbitant fee in Ukrainian hryvnia, you can slouch the night away on one of the sofas in the company of shop girls, actuaries and mutilated war veterans, all of them mellowed by very poor quality Afghan marijuana. Things haven't changed there much since Herodotus reported that 'the Scythians . . . take some of this hemp-seed, and, creeping under the felt coverings, throw it upon the red-hot stones; immediately it smokes, and gives out such a vapour as no Grecian vapour-bath can exceed; the Scyths, delighted, shout for joy'.[28]

Then there was opium, known for millennia in the eastern Mediterranean, cannabis, henbane (consecrated to Jupiter in the Roman world, to Thor in the Nordic world, and a key ingredient of the 'flying ointments' beloved of European witches) and, of course, alcohol. Euripides' description in *The Bacchae* of the effect of alcohol is echoed (although nowhere else as well) in every drinking song that has ever been written:

> God of merriment and festal crowns,
> He lightens our feet and our heart
> To honour him in the dance.
> The music of our laughter
> And the wild piping of flutes
> Are a portion of his largesse.
> Sweetest of all . . . the oblivion he brings,
> The respite from all care
> When the bright blood of the grape
> Bursts over the table of the gods,
> And ivy-crowned revellers
> Lift and tip the brimming bowl
> To drink warm, deep sleep. . .
> Take me, Bromios [Dionysus], lord of the dance.
> Bromios, god of bliss,
> Take me there,
> Where the Graces dwell,
> Where desire leaves its banks,
> Where the very air
> Sanctions our devotions and our dance
> To you.[29]

In mid-winter every other year, the maenads (female worshippers of Dionysus) processed out of the city shouting 'to the mountains'. In the dark of some sacred grove or high place, the music of the *aulos* and the *tympanum* struck up and pitch torches were lit. The maenads shook out their hair, pulled up their fawn skin dresses and began to dance. They jumped, whirled, ran

*The triumphal cortege of Dionysus, the Greek god of wine
and intoxication. He was also (and perhaps not paradoxically)
associated with the realm of the dead. Possibly the association
was a statement about the use of intoxicants as vehicles to
other realms. When the Greeks had happy thoughts about the
afterlife, the thoughts were often of a life blessed with wild
Dionysiac joys. (From a Tunisian mosaic, c. AD 200)*

and shook their heads until the frenzy took them. They fell to
the ground, foaming in ecstatic union with the god.

This is all very colourful, but is it significant? Does it mean
any more than that drugs do strange things to people – a proposition that needs no anthropological footnotes for credibility?

Can drugs be doorways to the sort of mystical experiences
that we read about in the biographies of the unimpeachable
mystics?

Some have thought so. Aldous Huxley thought that drunkenness was the poor man's symphony orchestra; a vehicle of sublimity and the numinous; a respectable suburb of the mystical consciousness.[30]

It is not only Huxley who waxes mystical, and not only alcohol that stirs up mystical language. Thomas De Quincey took opium for headaches. An hour later he was in bliss:

> What a resurrection from the lower depths of the inner spirit! What an apocalypse! . . . This negative effect [of the elimination of his headache] was swallowed up in the immensity of those positive effects which had opened up before me, in the abyss of divine enjoyment thus suddenly revealed . . . Here was the secret of happiness, about which the philosophers had disputed for so many ages, at once discovered.[31,32]

Similar extravagances can be found in most of the writing from the drug world, which has become a genre all of its own. LSD

Maenads and plainly aroused satyrs gather in the grapes for wine-making. (Attic black-figure cup, sixth century BC, Luynes Collection)

has caused people's consciousness to explode out of their heads; to turn them into the cosmos; to relive past lives; to become mythological creatures; to turn into wolves; to salvage long-submerged memories from early infancy and the womb, and transmute them into winged demons and armchairs; to do in an hour what a lifetime on the couch of a Freudian or Jungian psychotherapist could do. It makes Dali dull. As the mundane glitters dazzlingly with psychedelic jewels, so abstracts become concrete, grow wings, put their hands in their pockets and saunter round the room. Ultimate things are the basic furnishings of the entheogenic house.[33] Subjects may see their own death with dispassion, as if noting a fly buzzing out of the room into a summer garden. Sometimes, indeed, they might experience it.

The icon of the 1960s drug culture, Timothy Leary, described his experiences on ketamine as 'an experiment in voluntary death'.[34] Ketamine puts the taker 'in a state where he is completely divorced from his body and is projected into some astral plane as pure consciousness', wrote another user. 'Doing K this way is as close as he can come to dying without dying, and it is always a beautiful, wondrous experience.'[35]

I wonder how much this user really knew about ketamine. It is not always beautiful. But mystical imagery insistently peppers the ketamine blogs:

Well, doc, I was floating up on the ceiling, looking down at myself.

I died and went to heaven. I saw God and the angels. It was beautiful.

This UFO landed and these aliens took me aboard and we flew around.

Everything was red. Then I realized I'd become a blood cell.

I relived a previous life as an Egyptian mathematician.[36]

It does seem like theology at the end of a needle.

William James, who used nitrous oxide as a mind-altering substance, wrote:

> Looking back on my own experiences [with nitrous oxide], they all converge towards a kind of insight to which I cannot help ascribing some metaphysical significance. The keynote of it is invariably a reconciliation. It is as if the opposites of the world, whose contradictoriness and conflict make all our difficulties and troubles, were melted into unity . . . This is a dark saying, I know, when thus expressed in terms of common logic, but I cannot wholly escape from its authority. I feel as if it must mean something, something like what the Hegelian philosophy means, if one could only lay hold of it more clearly. Those who have ears to hear, let them hear; to me the living sense of its reality only comes in the artificial mystic state of mind.[37]

Two things. First, James notes what Graham Hancock and many others have noted – that the experiences on chemical hallucinogens seem *more* real than everyday experiences, not less. The drug world is super-heavy, super-hard, super-sharp. And second, 'I feel as if it might mean something' – the impulse to explain. Those two feelings have, over millennia, led to a desire to integrate drug-fuelled experiences into religion – to systematise and doctrinise their insights. They may have been the catalyst for some religions. Some have even suggested (a suggestion to which we come in detail in Chapter 11) that the religious impulse itself was triggered by ingesting or inhaling some psychoactive substance.

We have seen already, in the citation from the *Rig Veda*, how the insights given by the soma plant (whatever that may be) are cherished and celebrated in ancient Indian religion. We don't know whether soma was an architect of Hinduism, or was only a respected bard who told stories in Hinduism's house, but we do know of the centrality of iboga to Bwiti, of ayahuasca to

Peruvian shamanism, of peyote to many Central and North American indigenous religions, and so on.

Cannabis was loved by Shiva, who sat up in the high Himalaya smoking it and letting his insight trickle down the Ganges.[38] He is particularly pleased when hemp burns in his temples. And cannabis, of course, is a sacrament in Rastafarianism. References to 'herbs' in Genesis, Proverbs and Psalms are, say the Rastafarians, references to the weed itself.[39] There are even some scholarly supporters of the notion that the holy anointing oil described in the book of Exodus[40] and subsequently used in the Jerusalem temple contained cannabis. To describe Jesus as 'the anointed one', some contend, must mean that the sacramental hemp oil was physically applied to him.[41] Exegetically it's a hard thesis to get up and running, but once it's up it can take you to just about any weird theological address you like.

The Rastafarians are not the only ones who see hallucinogens in the Old Testament. Yes, there are the spectacular speculations (Ezekiel was on magic mushrooms, and so on: theses we have touched on already, and will meet again), but, more fundamentally, there is the suggestion, by R. Gordon Wasson and others, that the fruit of the knowledge of good and evil, so disastrously plucked by Eve in the Garden of Eden and so foolishly eaten by the weak Adam, was hallucinogenic. After the fruit was eaten, after all, 'the eyes of both were opened'.[42]

The fruit, for Wasson, was an *Amanita* mushroom, one of the psilocybins. And God himself, said Wasson, lured Eve towards it:

> Of arresting interest is the attitude of the redactors of Genesis toward the Fruit of the Tree. Yahweh deliberately leads Adam and Eve into temptation by placing in front of them, in the very middle of the Garden, the Tree with its Fruit. But Yahweh was not satisfied: he takes special pains to explain to his creatures that theirs will be the gift of knowledge if, against his express

The fly agaric mushroom, Amanita muscaria. A psychoactive mushroom used widely in Siberian shamanism, and probably by European and other Asian shamans too.

wishes, they eat of it. The penalty for eating it (and for thereby commanding wisdom or education) is surely death. He knew the beings he had created, with their questing intelligence. There could be no doubt about the issue. Yahweh must have been secretly proud of his children for having the courage to choose the path of high tragedy for themselves and their seed, rather than serve out their lifetimes as docile dunces. This is evidenced by his prompt remission of the death penalty.[43,44]

In a chapel at Plaincourault, Indre, France, there is a curious fresco, dated to 1291. It depicts the fall of man. The tree of the knowledge of good and evil is at the centre. But it is an odd tree: it looks very like a tall, branching mushroom with a spotted cap, just like the fly agaric. The serpent winds round the trunk or stem, offering the fruit to Eve. Eve herself has a pained face. Her hands cover her abdomen. Lines up and down her torso might indicate that she is bent over. If you want to see her as retching with the nausea that often comes from ingesting *Amanita muscaria*, it is not hard.

The thirteenth-century fresco of the fall of man, Plaincourault, Indre, France. According to John Allegro and others, this represents a 'mushroom tree', and indicates that magic mushrooms or other entheogens are responsible for the origin of both Judaism and Christianity.

If the biblical scholar John Allegro is right,[45] the Plaincourault fresco illustrates a late memory of a revolutionary truth – that hallucinogens were not only involved in the genesis of Judaism and Christianity, but induced the experiences that, codified, *were* Judaism and Christianity. Others, notably R. Gordon Wasson, had made this claim for Judaism, but Allegro provoked outrage by saying that he could demonstrate etymologically that Jesus did not exist, and was in fact the personification of a magic mushroom. His claims were not made any more credible by being published in a popular paperback and serialised in the *Sunday Mirror*, instead of being subjected to the usual academic peer review. Here, in the newspaper version, he summarises his contentions:

> From the earliest times the folk-tales of the ancients had contained myths based upon the personification of plants and trees. They were invested with human faculties and qualities and their names and physical characteristics were applied to the heroes and heroines of the stories. Some of these were just tales spun for entertainment, others were political parables . . . Here, then, was the literary device to spread occult knowledge to the faithful . . . Thus, should the talk [about the mushroom cult] fall into Roman hands, even their mortal enemies might be deceived and not probe further into the activities of the mystery cults within their territories.
>
> What eventually took its place was a travesty of the real thing, a mockery of the drug's power to raise men to heaven and give them the longed-for glimpse of God. . .
>
> . . .the story of Jesus and his friends was intended to deceive the enemies of the sect, Jews and Romans. It was a hoax, the greatest in history. Unfortunately it misfired. The Jews and the Romans were not taken in, but the immediate successors of the first 'Christians' (users of the 'Christus', the sacred mushroom) were. The Church made the basis of its theology a legend revolving around a man crucified and resurrected – who never, in fact, existed.[46]

Christianity, then, is a comic-tragic farce: a misunderstanding straightforwardly in the line of *Life of Brian*. Instead of a man called Brian getting mistaken for Jesus, the codeword 'Christ' (known to the initiates to mean 'hallucinogenic mushroom') is taken to mean a real person called Jesus Christ. Everyone is in on the joke except the humourless, dim, second-generation Christians who take the whole thing seriously and disseminate the joke, at colossal personal cost and with considerable theological embellishment, throughout the Mediterranean. The people who grasp the wrong end of the theo-fungal stick so firmly soon become the Christian establishment. When those who really know the truth (the Gnostics – of course – and others) try to remind the fledgling church what 'Christus' really meant, they are brutally suppressed – as, indeed, is the use of psychoactive mushrooms itself. Folk memories persist of a psychoactive sacrament that could give eternal life, but the priests of the blundering new orthodoxy substitute a placebo sacrament – the Eucharist. Echoes of the original meaning can still be heard in Christian prayers. We are solemnly assured that the bread and wine are the body and blood of God: the body of the original mushroom was the body of the God-mushroom, codenamed 'Christ' to avoid trouble.

Allegro is vague about the fate of the mushroom cult. He says that it lingered on for a little while. But Plaincourault hardly fits his thesis. Twelve hundred years is a long time. Perhaps, if pressed, he would suggest that the cult's fate ran parallel to that of Gnosticism – an over-discussed and under-evidenced subject if ever there were one.

It is harsh to say that Allegro's thesis is demonstrable nonsense. But we can get fairly close. His etymological arguments have received short academic shrift, and he really has little else. Every bearing that we can get on Jesus places him firmly in a human body, in first-century Palestine, in the Jewish tradition of apocalyptic prophets, and nowhere near the fungal kingdom. And there would have been no need for the mushroom, in order to maintain its cover, to have preached the Sermon on the Mount. Allegro can point out that Plaincourault is by no means the only 'mushroom tree' in medieval art,[47] but the consensus amongst art historians is that they represent highly stylised versions of other trees – possibly Italian pine trees.[48]

That consensus, to be fair, is by no means fatal to Allegro. He might well ask, 'Why stylise like *that*?' But the mere form of the tree takes nobody much further.

Allegro can draw some very limited support from Wasson's contention (itself terribly weak: an assertion, not an evidenced argument) that the Eden story depicts hallucinogens, by pointing out that at the very end of the Bible, in Revelation 2 and 22, there is another tree.[49] 'There it is again,' goes the argument. 'This indicates that the *true* early church knew exactly what the *real* Judaeo-Christian revelation was all about: it was about what you see when you've had psilocybin mushrooms.' The obvious difficulty in enlisting Revelation on the side of the mushroom is that the Revelation tree is the tree of life, not the tree of the knowledge of good and evil.

There have been some other ingenious attempts to see parallels between the supposed references to psychoactive mushrooms in the Old Testament and those in the New. Clark Heinrich, for instance, cites Ezekiel:

> I looked, and a hand was stretched out to me, and a written scroll was in it. He spread it before me; it had writing on the front and on the back, and written on it were words of lamentation and mourning and woe. He said to me, O mortal, eat what is offered to you; eat this scroll, and go, speak to the house of Israel. So I opened my mouth, and he gave me the scroll to eat. He said to me, Mortal, eat this scroll that I give you and fill your stomach with it. Then I ate it; and in my mouth it was as sweet as honey. He said to me: Mortal, go to the house of Israel and speak my very words to them.[50]

The references to 'lament and mourning and woe', argues Heinrich, are plainly to be associated with the eating of the scroll. They are references to the bodily pains you get if you eat fly agaric. Fly agaric caps are pliable, can be scrolled up, and when the upper layer of cells is peeled off, the cap can look as if it is inscribed with cryptic writing. They also have a sweet, honey-like smell. Enlightened by what he has seen in his *Amanita*-induced visions, the prophet is equipped to speak to the house of Israel the words he has received from God (a.k.a. the mushroom).[51]

As in Ezekiel, so in Revelation. John of Patmos describes a magic mushroom session.

> Then the voice which I had heard from heaven spoke to me again, saying, 'Go, take the scroll which is open in the hand of the angel who is standing on the sea and on the land.' So I went to the angel and told him to give me the little scroll; and he said to me, 'Take it and eat; it will be bitter to your stomach, but sweet as honey in your mouth.' And I took the little scroll from the hand of the angel and ate it; it was sweet as honey in my mouth, but when I had eaten it my stomach was made bitter. And I was told, 'You must again prophesy about many peoples and nations and tongues and kings.'[52,53]

These passages, taken by themselves, do provide a slender foundation for the contentions rested on them by Allegro and Heinrich. But as soon as they are joined by the rest of the Old and New Testaments, the foundations collapse entirely. Look at what Ezekiel went on to prophesy about. There's no secret about what 'lament, mourning and woe' referred to, and it's not stomach ache or queasiness. Eating something is a clear reference to the commissioning of the prophet's lips. It is sweet because it is holy: it has come from a holy God. It is bitter, in Revelation, because it contains solemn words. The bittersweet scroll tells of the necessary coexistence of judgement and salvation.

Allegro et al. sound absurdly outrageous, and in many respects they are. But they do make some serious points, which have been unfortunately under-examined. It has not been academically reputable to take John Allegro seriously. It may be that there was some entheogen use in Judaism and early Christianity. It may be that some of the experiences on entheogens lent some images to Judaism, Christianity or, more likely, Gnosticism. So far there is no evidence at all that they did, but the subject should not be outlawed. It is worth a serious scholarly look. But what does seem reasonably certain is that if entheogens lent anything at all to Judaism or Christianity, they did not lend foundational experiences or any significant doctrines. They might conceivably have lent some artistic colour: they do not lie behind any comma, let alone any sentence, of a creed.

We're hardly any further on in answering the question, 'Are experiences on psychoactive drugs genuinely mystical?' Perhaps the question is meaningless because the word 'mystical' is so vague. Similarly for 'religious'. We know that drugs produce interesting visions. It would be surprising if mystical and religious tags were not slapped on those visions by the users, without necessarily wondering too hard about definitions.

There are undoubtedly some parallels between some of the things that are seen in drug-induced visions and those recorded in the great religious writings. If you want to see mushrooms at work in Ezekiel, it's easy enough. Wasson, who denied that psychoactive drugs were implicated in the Revelation visions, nonetheless observed that the literary flow of the book of Revelation is similar to the rolling ride in the car of your own consciousness that you get on magic mushrooms.

But here, perhaps, is the point. You don't need to take exogenous drugs to have, pharmacologically, the effect of those exogenous drugs. Most of the main psychoactive drugs are either analogues of naturally occurring neurotransmitters, or change the levels of naturally occurring neurotransmitters. As any fasting ascetic or dancing shaman will tell you, you needn't pop pills to enter other states of consciousness. But the trance of an exhausted, dehydrated shaman might be pharmacologically similar to the ayahuasca trip of a well-fed Californian tourist. There's also the issue of susceptibility: some neurones might shoot off mystical fireworks at lower levels of a particular natural neurotransmitter than others.

Aldous Huxley – a man notoriously careful with his labels, and well versed in the literature of comparative religion – was in no doubt that drugs could deliver the real religious article. Just like the standard mortifications of fasting, sleeplessness and masochism, they could change body chemistry in a way that could fling open the same doors of enlightenment through which the great saints and mystics dance for their audience with ecstasy:

> . . . a person who takes LSD or mescaline may suddenly under-
> stand not only intellectually but organically, experientially, the
> meaning of such tremendous religious affirmations as 'God is
> love' or 'Though He slay me, yet will I trust in Him. . .'[54]

Let's suppose, for a moment, that the foundational contentions of (say) Christianity are true, and that it is possible (for instance) for the Holy Spirit – the Spirit that devised the equations of the Big Bang, flung the stars into space, and devised the human mind – to come in some circumstances to dwell within a human being. One might expect that advent to have some very dramatic neurological sequelae. It would be surprising if our dopamine and serotonin levels remained unchanged and our thalamus retained the sort of respectable equilibrium it is in when we're sitting at our desks. It would be very strange if things didn't go joyfully haywire. They might well go haywire using some of the pharmacological pathways travelled on an LSD trip.

But if the *theological* contentions of Christianity are true, the haywireness won't be random. The pathways will tend to go in particular directions. They will tend to produce the sensation of an intimate personal encounter with a totally accepting Being. It won't just be an encounter with an anonymous Power. Christianity is not about the Force being with you, but about the Person being with you. The sensations will have some sort of ethical colour: they will tend to make the subject a better person. And they will have the sort of consistency that is generally lacking in drug-induced epiphanies.

The main difficulty in writing this chapter has been that of doing justice to the immense variety of drug-induced experience. And of course I have failed. But that failure is significant. On drugs you go into a huge number of worlds and meet a huge number of beings. Some beings are benevolent, some malignant, some ambiguous and capricious. Very often there are no meetings at all. Some worlds (such as the ayahuasca world) are lush and vegetative. Others (such as the pure dimethyltryptamine world) are hard, geometric and artificial – the world of *The Matrix*. Sometimes there is bliss; sometimes, on the same dose and for no obvious reason, there is terror. The drug world, in short, is immensely various and immensely inconsistent.

The experience that has been regarded as legitimately Christian is not quite as varied, and nothing like as inconsistent. There is always a sense of security. You go into a safe place. You encounter a Person – or at least there is an overwhelming impression of personality. There is often bliss. Sometimes the bliss is so intense that metaphors from sexual ecstasy and suffering have to be deployed to describe it. There is often awe, mingling with an essential homeliness, but never horror. There is never a 'bad trip'. The worst thing that happens is a painful realisation of one's shortcomings, but it is a realisation that comes simultaneously with a knowledge that mercy and patience swamp justice. And it often changes lives. 'Now that I've met you,' goes a popular Christian song, 'I'll never be the same again.' And it seems often to be true.

The well-known introduction to Christianity, the Alpha Course, is fond of testimonies about how an experience of 'the Holy Spirit' on the course has produced moral rehabilitation. A typical story in the Alpha promotional literature will recount how a hardened jailbird with a broken marriage was prayed for on the Alpha weekend. He will 'ask the Holy Spirit to come into his life', and will feel some sort of overwhelming sensation of joy. Tears will well up. He may, perhaps slightly later in the same weekend away, receive the 'gift of tongues' – *glossolalia*. He will describe how he feels love for everyone: how, from being a disruptive, violent prisoner, he became a friend to all, obtained early release from his sentence, put his marriage back together, got a job, became a pillar of the local church and is happy, fulfilled and (that word again) joyful. He's on a trip that lasted.

But drugs can also have apparently good, lasting effects. If 'by your fruits you shall know them', there is some evidence that some of the psychedelics might be a Good Thing. Sometimes unimpeachably correct ethical insights emerge from drug trips. Ibogaine gave a heroin addict a salutary vision of what would happen if he didn't kick his habit. 'Clean up your room!' he was told. 'Ibogaine', he said later, 'is God's way of saying: "You're mine."'[55]

On Good Friday in 1962, a group of twenty male Protestant theology students met in the basement of Marsh Chapel at Boston University. In a double-blinded trial[56] devised by the Harvard psychologist Walter Pahnke, half were given a capsule containing psilocybin – the psychoactive element in psilocybin mushrooms.

The others were given an 'active placebo', niacin, which produces some physical but no significant psychoactive effects. Above them in the chapel, a Good Friday service was in progress. It was piped down to them. One of the recipients of psilocybin, Mike Young, who later became the minister of the Unitarian Universalist Church in Tampa, described what happened.

'I just slid into it very gently, very, very beautifully,' he said. 'Colors became incredibly intense. Geometric figures seemed to etch themselves around objects. When somebody moved there was an after-image, a flare behind the motion.' He closed his eyes and 'leapt into an incredible kaleidoscope of visual wonderment'. Music, poetry, scripture readings and a sermon from the service crept into his consciousness. He went to urinate. Cigarette ashes in the urinal looked like black pearls. He heard cars racing by outside. 'I didn't know which was the real world. I couldn't keep straight what was happening inside my head and what was happening outside.' He returned to the room, and was hurled into 'the major vision of the drug for me . . . I was awash in a sea of color. These bands of swooshing liquid. It was like being underwater in an ocean of different color bands. Sometimes, it would resolve into patterns with meaning, and other times it would just be this beautiful swirl of color. It was by turns threatening and awe-inspiring.' Slowly the colours crystallised into patterns. 'It was a radial design, like a mandala, with the colors in the center leading out to the sides, each one a different color and pattern.' He felt as if he were a fly at the centre of this huge psychedelic web. 'I could see that each color band was a different life experience. A different path to take. And I was in the center where they all started. I could choose any path I wanted. It was incredible freedom . . . but I had to choose one. To stay in the center was to die.' But the choice was agonising and impossible. 'I couldn't choose. I just . . . couldn't . . . pick one.' He felt that he was being disembowelled. 'It was incredibly painful.' Paralysed by terror and indecision, he waited, his guts churning. 'And then', he said, 'I died.'

Over the PA system, the preacher read out an Edna St Vincent Millay poem: 'I shall die, but that is all I shall do for Death. I am not on his pay-roll.' Young scrawled something on a piece of paper. Later he looked at it. He had written, 'Nobody should have to go through this. Ever!' 'I wasn't talking about the drug trip,' he explained. 'I was talking about having to make this choice of what to be. I was talking about having an ego and having to have it die in order to live in freedom. I had to die in order to become who I could be. I did make a choice, in that willingness to die.'

Young's trip lasted for another three hours. 'I was in and out of vision, but it was pleasant. Interesting. It gradually tapered off. I was gently coming down and reflecting back on that death image.' Eventually the world came back into focus. Along with the other subjects he went off to Timothy Leary's house for sandwiches.[57]

Similar stories came from many of the psilocybin subjects. In questionnaires administered shortly after the trial, they all gave their experience far higher ratings for various mystical qualities than did the niacin subjects. But perhaps the more interesting results came later. When assessed six months after the study, the psilocybin group reported that they had become better people. They were more convinced of the truth of their faith, more loving, sensitive and considerate, and rejoiced more in their lives than they had done before. The pro-drug lobby celebrated: here, they said, is impeccable, Harvard-endorsed proof that hallucinogenic drug users are good citizens. They should be applauded, not locked up.

Twenty-five years later, Rick Doblin, also at Harvard, cracked the codes preserving the anonymity of the subjects in the Good Friday experiment and interviewed many of them. He reported:

> In the long-term follow-up even more than in the six-month follow-up, the [psilocybin] group has higher scores than the control group in every category [for perceived positive effects from the experience] . . . A relatively high degree of persisting positive changes were reported by the experimental group while virtually no persisting positive changes were reported by the control group . . . All psilocybin subjects participating in the long-term follow-up, but none of the controls, still considered their experience to have had genuinely mystical elements and to have made a uniquely valuable contribution to their spiritual lives.[58]

Of course Doblin could not know if the psilocybin group were more altruistic than they would otherwise have been, beat their wives less than the control group, and so on. But the results are interesting as far as they go. It is a scandal that such important, potentially far-reaching work was not repeated and that we have to rely on a very small study from 1962. But the dangerously narrow-minded Federal Drugs Administration in the United States outlawed all scientific research on hallucinogens until very recently.[59]

Doblin also criticised the methodology of Pahnke's trial. He was scathing about Pahnke's failure to mention that one of the psilocybin subjects needed to be chemically tranquillised during the experiment, and about Pahnke's underplaying of the psychological struggles of most of the psilocybin subjects to come to terms with their experiences (albeit that in the end the experience was perceived as a positive one). Those criticisms are significant. They make again the point that there is a relatively high incidence of bad trips (even in a controlled, safe environment, in an explicitly religious, comforting context).

In a 2006 study of psilocybin at Johns Hopkins University, almost a quarter of subjects had significant fear, often associated with nasty paranoia.[60,61] When John Horgan went off to the Pacific coast of the US for an ayahuasca night, his host, Tony, told him that 'usually at least one person becomes convinced he's going insane or dying', and that they had borrowed shamanic techniques to deal with it.[62]

Psychiatrist Karl Jansen notes that ketamine experiences can sometimes be terrifying. He quotes one user:

> I went to Hell . . . I took the largest dose ever, 200mg as a shot in the buttock, and curled up for the night. Once again I was going through a pipe system, but this time I came out into a small, dimly lit glowing red room, and was filled with terror. I had lost my body and had become something hanging on a peg. I thought that I would have to stay forever in that room. I was in HELL. I screamed, and screamed . . . it was my first experience of what ETERNITY really means. I almost became a Catholic the next morning. That experience seriously shook the basis of my disbelief. I thought that people were absolutely mad who wanted to carry on for Eternity without their bodies. It made me fervently hope that the end really is the end.[63]

An enthusiastic ketamine user, seeing the soul as the entity that normally does effective battle with evil things when it is in our body, thinks that the real danger comes not from what you meet on the trip, but from the burglars who might visit when you're away. He warns ominously, in a drugs news group:

> You need to have absolute faith that your body-mind will take care of itself 100 per cent while you're away . . . Meditation helps, but I recommend spending lots of time in a float tank. 'I'm going now, body. Take care. Be back later.' When you return, the first thing you do is a body check: Heart? OK. Breathing? OK. Senses? Eyes – out of focus etc . . . Please note that when you are not using your body (hardware), other non-physical entities (software) might ask to use it.

Between 1990 and 1995 Rick Strassman, a psychiatrist in New Mexico, obtained permission for studies on sixty human volunteers using DMT – the active ingredient in ayahuasca. He suspected that mystical experiences might be produced by the secretion of endogenous DMT from the pineal gland. Some of the volunteers indeed reported the sort of mystical experiences written up in the religious classics. They floated timelessly in a sea of bliss, a sea of which they were a part. The boundaries between them and the world dissolved; they knew that their consciousness would survive for ever. Some drifted through the clichéd spiralling tunnel towards the bright, ineffable welcome.

But it was not good for everyone all the time. About half the subjects met weird creatures: elves, insects, alien-like humanoids, clowns, robots and beings so strange that all labels failed.[64] Some were friendly, some couldn't care less, and some were vile. Volunteers were threatened, terrified and subjected to agonising experiments. One man was anally raped by two dinosaur-like creatures. Huge insects ate another alive.

Strassman, to his profound embarrassment, found himself unable to say that these visions had been dredged up from the volunteers' subconscious, and finally postulated that DMT had propelled the subjects into a real dimension – one of the hyperspaces hypothesised by quantum theory. It didn't do much for his career. Whether he was right or not doesn't matter for our purposes. The important point is that the DMT world was alarming and at least partly bad. Strassman, badly shaken, abandoned his research.[65]

The casualties cannot be forgotten. Generally, in the hysterically uncritical literature proclaiming the mystical benefits of drugs, they have been.

The broad picture is this: some drugs can duplicate fairly accurately the experiences of drug-free highs reported by people in religious ecstasy. Some of those drug-induced experiences can have lasting effects – certainly in terms of perceived well-being. They are seen by the participants as being spiritual bequests of immense and sometimes supreme value. They may cause the budding of ethical fruit too, although the evidence for that is less convincing, and one would presumably, to weigh those fruits, have to take 360-degree soundings from a large sample of the people who had contact with the subject. Niceness and altruism are hard to assess objectively. A fair proportion of the drugged subjects, but no significant proportion of the non-drugged controls, had nasty, frightening experiences on the way.

It looks very much as if drugs work through some, at least, of the same pathways that are used in non-drug religious experiences. Isn't that what we'd expect? It would be surprising if there were neuronal tracts dedicated exclusively to God's use. Indeed, the possible existence of just such a mysterious conduit, implied by Newberg's imaging studies on *glossolalia*, rightly raises neurological and theological eyebrows.[66] The casualties on drugs – the bad trips – imply that the pathways aren't meant to be used that way. Drugs aren't a very good fit with our heads. It looks as if our brains are designed for religion, but not for drugs.

Does that mean that drugs have no place at all in religion? That the parroting by drugs of mystical sensations is a sort of blasphemous pharmacological parody? Well, not necessarily. Indeed, Jesus himself not only legitimised in religion the use of one fairly potent psychoactive agent – alcohol – but appeared to make it mandatory.

'Ecstasy [the drug MDMA] opens up a direct link between myself and God,' said a Benedictine monk. 'It has the capacity to put one on the direct path to divine union . . . It should not be used unless one is really searching for God.' He used ecstasy as a sort of spiritual dynamite to blast out of the way the blocks and distractions that prevent effective prayer. He only used it two or three times a year, but it had given him 'a very deep comprehension of divine passion'.[67] A Rinzai Zen monk gave ecstasy some of the credit for his promotion to abbot. It was most effective on the second day of a seven-day meditation, he had noticed, because it was then that there was the greatest danger of becoming distracted by bliss.[68] And at Harvard, a group of eight graduating seniors from the Divinity School used ecstasy as a sacrament in a ritual that they called the 'Harvard Agape'. 'It was an amazing grace,' reported one of them, 'that grace that passes understanding. I was moved; I was in communion with everyone else in the room. It was as if, at that moment, all barriers had come down, all suffering had ended, all pain had been relieved, all joys had been known. I forgave the offences I had suffered and was forgiven for my sins. I was healed. I was strengthened. I was redeemed.'[69]

This was very sloppy theological drafting by that Harvard Divinity student. He should have known better. What he meant was that he *knew* experientially, in a way that he had not known before, that he was forgiven, healed and redeemed. Now, that experiential knowledge is good: indeed, it is crucial. But if it could only be generated by ecstasy, there is something wrong with his theology, or the means by which his emotions connected with his theology, or both. In either case the lasting solution isn't a pill of E. Indeed, a pill of E, even if it always gives a good, theologically sound trip, is likely to give dangerously false reassurance. It won't correct his theology, and is unlikely in itself to forge a lasting connection between his books and his heart. It would be better if he learned how to be ecstatic using the heady cocktails of his own theology and his own neurotransmitters, and if the Rinzai Zen monk learned how to grapple with his bliss.

But what of Jesus and that strange eucharistic injunction? 'Take, eat, this is my body,' and, 'Drink from [the cup], all of you, for this is my blood of the covenant.'[70] It is strange indeed, and we return to it in Chapter 13.

Huxley thought that a religious renaissance would result from the use of entheogens, not from evangelistic crusades. Drugs would allow, in the comfort of the user's own home, a dramatic self-transcendence and a revolutionary knowledge of the nature of things. The renaissance would be more than simply a multiplication of proselytes. Religion itself would be changed.

> From being an activity mainly concerned with symbols, religion will be transformed into an activity concerned mainly with experience and intuition − an everyday mysticism underlying and giving significance to everyday rationality, everyday tasks and duties, everyday human relationships.[71]

Huxley has a low, common and confused view of what religion should be. 'Mainly concerned with symbols'? That's the talk of someone scarred for life by what the chaplain at Eton did to him. 'An everyday mysticism underlying and giving significance' to the everyday? Isn't that precisely what Christianity, Judaism, Islam, Hinduism and just about every other religion purports to be? But perhaps that's his point: they purport to be, but they are not. They talk the talk, but don't walk the walk. They let their practitioners look at a text describing the truth; they don't help them to taste the truth, to live within it, to let it tingle on their skins and gurgle in their guts. If that's his criticism, I'm with him almost all the way. To say that you can have religion without experience is as laughable as saying that you can love truly a woman you've never met.

I expect that everything, including the equations of Newton and Einstein, began with an experience. I doubt that anything at all was worked out entirely as a sense-free set of propositions, and am certain that nothing worthwhile was. I imagine that the genius of a truly ground-breaking mathematician is a form of synaesthesia, of the sort that you sometimes get on

DMT, when you taste numbers, or see sounds. Someone who tastes a number – who *experiences* it – knows more about it and understands it more deeply than someone who sees it on a piece of paper and is simply adroit at manipulating it. A religious person is one who has tasted God, not learned some facts about him. God doesn't taste of ecstasy: he is *the* ecstasy.

Lots of vehicles use our road in Oxford. Many of them are cars. They sound pretty much the same, to someone whose ear is as untuned to engines as mine. The other day I stood in the front room of our house waiting for my wife and children to return. I heard many cars going past. Each of them produced in my head an initial sensation identical to that which was eventually produced by our car when my wife finally drove up. But then the front door was unlocked, and the boys burst anarchically in, and my wife followed, and the sensation then was wholly different from that caused by the previous cars.

It is similar with drugs. LSD might use some of the biochemical roads, and produce in some people, some of the time, a few of the sensations associated with the movement of God, but the sensations it produces are no more evidence that it (or DMT, or mescaline, or opium, or anything else) is somehow to be *identified* with real God-induced sensations, than the similarity between the sound of the cars in our street is evidence that everyone driving those other cars was in fact my wife. You know the real thing when it arrives. If you look hard enough at the mass of evidence about drug-induced experience, there's little real danger of confusing or conflating the Tree of the Knowledge of Good and Evil, the Tree of Life and the Vine of the Dead.

CHAPTER 7

Finding God in the Bedroom:
The Sexuality of Spirituality and
the Spirituality of Sexuality

'Why are you celibate, Mr Hollaston?'
   'How do you know I am?'
   'Because your power is so great, though you know so little.'
                              (Geoffrey Household, *The Sending*,
                        London: Book Club Associates, 1980, p. 141)

'If you're writing a book about ecstasy,' said a slightly drunk
Professor of Mathematics, 'you'll have to meet Monica. Won't
he, darling?'

'Of course you will,' said his wife, 'but be careful, won't you?
Wear all your worst clothes and take a big stick.' And she growled
like a hunting lioness.

Monica, it seemed, was notorious in north Oxford as a practitioner of tantric sex – the Eastern practice of ancient but
obscure origins, with probable roots in Hinduism, Jainism,
Taoism and Buddhism, which eschews frictional orgasm, relying
instead on a prolonged interlocking to catapult both partners
to simultaneous spiritual and sexual summits. From those
summits, it is said, one sees oneself, one's partner and the universe
as never before.

With some misgivings, and a very red face, I phoned Monica.
Could we have a chat? She would be delighted, she said, and
so we met up in a café that I knew would be empty.

She was late, and so I sat there drinking tea, wondering

what on earth I was doing, and reading about tantrism – the whole religious system of which sexual practices are only a small part. Tantrism's object is freedom – liberation from the grinding cycle of death and rebirth, from the ignorance that keeps us bound to the wheel, from the grey valleys through which we plod. Tantra, wrote David Gordon White, is the corpus of beliefs that 'working from the principle that the universe we experience is nothing other than the concrete manifestation of the divine energy of the Godhead that creates and maintains that universe, seeks to ritually appropriate and channel that energy, within the human microcosm, in creative and emancipated ways'.[1] The tantric literature oozes with this sort of stuff. 'Narrow it down a bit!' I'm always screaming. But they rarely do.

Hard-core tantrists, I read, are often sniffily dismissive of Western pop-tantra, with all its emphasis on the sexual rites. And certainly lots of the material I had read about tantric sex (covertly in bookshops I never normally go to) seemed to do nothing more than wax poetic about the ordinary sexual act. I'm all for celebrating sex, but that sort of celebration didn't seem to me to need the vast theoretic underpinning of the tantric movement, all with Sanskrit footnotes.

And then Monica arrived, gushingly apologetic for being late. I'd imagined her as a musky, kaftan-clad harpie with dark lascivious eyes, the painted nails of a predator, and a black book full of married men's mobile numbers. But she was in her early fifties, with grey sensible hair, a grey sensible suit and low sensible shoes. She carried a battered briefcase full of books on tantra, all marked up with yellow stickers. She had been married for thirty years to Alan, an antiquarian bookseller who played bowls and grew prize-winning leeks. She baked cakes, ran jumble sales, took the collection at an Anglo-Saxon church of ravishing beauty and minute congregation, and was the secretary of a charity that put up nest boxes in the local woods. Their numerous children had all proceeded effortlessly to her husband's old college in Cambridge and were now doing good works in hot places. I was charmed and rather disappointed.

'You'll think', she said, after she had settled down with a flapjack and a decaff coffee ('Nothing *too* strong at this time in the afternoon'), 'that I'm a deviant. A crazed sex-addict. That's what they all think.'

I hastily and dishonestly assured her that that was the last thing I thought.

'Well, of course you do. And I can understand it. Sex, for almost everyone here, is so *dowdy* – so dull, so momentary. A quick spurt and then to sleep.'

I looked nervously round to see who was listening. A pale student was watching us over the top of *Scientific American*. He looked quickly back to his article on quasars.

'They think of me as some sort of Ambassador for Eroticism,' Monica went on. 'And they giggle. I know they do. But I like to think of myself as running a Campaign for Real Sex.'

The student choked on his scone.

'Ordinary sex', she said, 'is so degrading for both partners. It's just a sort of masturbation which happens to involve someone else. It doesn't involve anything about either person apart from the body. And although bodies are wonderful, they are far more exciting if you can realise how *connected* they are to everything else. Making love to a whole person is so much more thrilling than making love simply to their body.'

So far, so conventional. This might have been a lecture to teenagers about being sexually sensitive and respectful. But she was just warming up.

'I think the secret is to realise the connectedness of the other person to everything else in the universe. To know that they are built of the dust from stars; are closely related to field mice;

were squeezed into the light through a tunnel of flesh; are part of a nexus of people you will never meet; to know that they will die. It is far easier to recognise these things in another person than in yourself. To think of yourself as a creature of the stars seems like psychosis or narcissism. And although we can believe that our lover will die, we can never really believe it of ourselves. But if we acknowledge all these things in another, and then become one with that other, they are somehow our bridge to all the other things of which we can see that they are a part. It establishes my communion with stars, field mice, and also with death. Because I know that Alan will die, and I become ecstatically one with him, I am somehow reconciled to my own death. The orgasm draws the sting.'

The student had given up all pretence of interest in quasars.

'But why is the orgasm important?' I wanted to know. 'Why isn't friendly solidarity enough?'

'It might be,' replied Monica, 'for a rare, advanced saint. I think.' And then she tailed off, and looked puzzled.

Nothing in life has prepared me, or I hope ever will prepare me, to ask a middle-aged pillar of the Women's Institute what she does between the sheets. But I had a pretty good idea. After some ritual washing, perhaps the lighting of a joss stick, and a mumbled incantation to the Vedic goddess Shakti, the couple would embrace and copulate. They would lie enlocked, for all the world like earthworms on a path (I could not dismiss the unfortunate image), perhaps for hours. Slowly the boundaries separating them would seem to dissolve; their skins would get thinner; and finally they would flow into each other and also, at the same time, into the world beyond themselves. On a good day they would taste the salt smack of *advaita* – non-duality – on the lips of the other. The climax would not merely be in their genitals, but in their noses, their toes, the bedside table and the far side of the sun. I knew that the neurophysiologists had been stalking Monica, but wasn't hugely impressed with their results.[2]

'Do you worship Shakti?' I asked. She suddenly looked more confident. She looked brightly up from her cappuccino. 'My dear boy,' she said, 'of course I do. And so do we all.'

A few months before I met Monica I had wandered through some of the great temples of south India, looking at some of the exuberant erotic carvings. At one point I had the misfortune to get snarled up with a Christian tour group from Mississippi. They were all over sixty and wore identical red baseball caps tattooed with some biblical reference in the King James Version. They had come, it seemed, to pray against the demonic spirits that had commissioned these sculptures and were still apparently inherent in them. They stopped in front of one of the finest frescoes, pursing their lips in disgust as Shiva pleasured his consort.

'Let's direct a beam of faith towards it,' thundered the leader. They stretched out their arms martially towards the abomination. 'Lord God,' continued the leader, 'as you raised your mighty hand to slay those who opposed your people, we ask you to break the power of uncleanness that defiles this place, and let your Word spring up to bring purity and healing.'

I watched, wondering what they hoped to achieve. Did they expect the whole magnificent edifice to crumble like the walls of Jericho? Did they expect Shiva to brush down his dhoti and start distributing Bible-reading notes?

I nudged one of the friendlier-looking women. 'What's all this about?'

She opened her eyes, put down her arm, and turned to me, irritated. 'What's it all about? It's about reclaiming the kingdom of darkness for the Light.' She closed her eyes again and joined in the praying, as if there was nothing else to be said.

I nudged her again. 'But why pray here? In front of these particular statues?'

She took a deep breath. 'Well, it's filthy, isn't it? It shows the depths from which they have to be delivered.'

'Would you cut the Song of Songs out of the Bible?'

'Of course not,' she hissed. 'That's all about the love of God for his people. And now, if you'll excuse me. . .' And she turned again to zap Shiva.

I saw them later, climbing into their air-conditioned bus with its witty sign, 'Tamil Nadu rains, but Jesus reigns', en route to the McDonald's franchise in Chennai. They sped off down the rutted road singing 'He's got the whole world in his hands'. With a slapstick scream a thin cyclist swerved into a ditch to avoid annihilation. Or, if the singers were right, worse.

That night, in my doss house, I turned up the Song of Songs, the most beautiful celebration of physical love from the ancient world. For sheer erotic power it makes the poems of Sappho and Catullus seem like gauche adolescent fumblings. I don't buy for a moment the notion that it is *merely* an extended metaphor about the love affair between Christ and the Church. The men who decided that it deserved a place in the Bible between the world-weary preacher of Ecclesiastes and the great herald Isaiah knew better than that. Yes, it no doubt works on that level too. Yes, it is a truth universally acknowledged that all really moving charismatic worship songs must have a breathy, panting female backing singer as the song reaches the climax of its adoration. But the Song wouldn't work on that level if its more immediate context – the heat of the bedroom – weren't recognised as a holy thing.

And yet I sometimes feel as if the Song of Songs is just *too* good. It lets the Judaeo-Christian tradition say, 'Yes, we do sex rather well, actually: just turn to the Song.' One of the unfortunate by-products of that is that sexual ecstasy is hermetically penned into the book. It doesn't seep out much into the rest of the Bible. And it is *so* wonderful that it sets the bar very high. There's a tendency, in the modern suburban bedroom, to think that you can't possibly say to your lover, 'Your two breasts are like two fawns, twins of a gazelle that feed among the lilies,'[3] and so there's no point in trying.

It's not that the rest of the Bible is sex-free: far from it. There's lots of libido sloshing around, particularly in the Old Testament, but it's often nasty and exploitative, and leads to terrible things. Sex in the New Testament is overshadowed by Paul, who was too busy raging about the sexual immorality of Corinth to give very measured advice and anyway thought, wrongly, that Jesus' return was so imminent that there was no time to smooch.

Although sex is celebrated in the Bible, the celebration takes two forms. It is good for loving couples to do this, it says. And, 'Imagine the best, most committed, most passionate relationship that two lovers can have. Got it? Well, God loves you far more than that.' There's nothing akin to tantrism. There's no suggestion of an idea that sex takes you into an altered state of consciousness in which you have specifically religious insights. A Judaeo-Christian ejaculation won't squirt your consciousness into astral projection. A Jew or a Christian makes love and then turns over and reflects theologically on the great simile in the Song of Songs.

Other cultures, to which Israel was opposed, apparently did see sex as a specifically religious act. There are plenty of temple prostitutes amongst the benighted heathen of the Near East, and they are all stoutly denounced by the stern prophets of Israel. Of course it is primarily the whole religious system of which the prostitutes were a part that was denounced, but it is hard not to hear some icy, puritanical disapproval of the sexual element itself. The whole raft of laws about sexual cleanness[4] and the very fact that Yahweh chose circumcision as the defining mark of his people indicate both an acknowledgement of the potency of sex and a caution about its deployment.

Sex as a direct conduit to some deity or some other world is recorded, but it is unusual. It has been used to consolidate (the kindest word I can use) the guru-devotee relationship. Bhagwan Shree Rajneesh infamously exhorted promiscuous sexual activity between members of his cult, but hardly gave a coherent account of why he was doing so. Witchcraft, both white and black, sometimes uses sex as a sacrament, but far more common than ritual copulations with Satan are pretty ordinary couplings under a summer tree in which the couple show their solidarity with the fecund nature around them by imitating the birds and the bees. The prominent erect phallus in the rock art of the southern African bushmen indicates that sexual excitement was a by-product of the shamanic trance depicted there, but sexuality does not seem to be a central pillar of shamanic ritual there or generally.

Bernini's explicitly erotic representation of the divine ecstasy of Teresa of Avila. (1652, Santa Maria della Vittoria, Rome)

Abstinence from sex, though, is ubiquitous in religious practice worldwide. Many of the explicitly erotic references in the Christian tradition appear in the context of sex expressly sublimated; erotic energy diverted away from the bedroom towards the altar. The most famous example is Teresa of Avila: we meet her in Chapter 11. But there are many others. In St John of the Cross's *Spiritual Canticle* the desolate bride (for which read the soul) has lost the bridegroom (Christ) and hunts long for him. They are finally reunited, with great joy:

The bride has entered
The pleasant and desirable garden,
And there reposes to her heart's content;
Her neck reclining
On the sweet arms of the Beloved.
In the inner cellar
Of my Beloved have I drunk. . .
There He gave me His breasts,
There He taught me the science full of sweetness.
And there I gave to Him
Myself without reserve;
There I promised to be His bride.
We shall go at once
To the deep caverns of the rock
Which are all secret,
There we shall enter in
And taste of the new wine of the pomegranate.
There You will show me
That which my soul desired;
And there You will give at once,
O You, my life!
That which You gave me the other day.
The breathing of the air,
The song of the sweet nightingale,

The grove and its beauty
In the serene night,
With the flame that consumes, and gives no pains.[5]

God and the flesh are of course seen as enemies in monasticism generally. Beat the body into submission by fasting and other mortifications, and you will somehow be more acceptable to God. One of the ways in which that acceptability is shown is by an intensification of religious experience. We will see in Chapter 9 that some of the traditional ascetic practices are themselves triggers for altered states of consciousness, but often, too, the non-sexual mortifications are seen as useful in reining in the rebellious libido. In lots of religious writing the primary enemy – the primordial dragon writhing in the subconscious deep, implacably opposed to God – is sex. Standing in cold water, staying awake when exhausted and fasting until you drop are seen as weapons in the war against the dragon. It's the dragon you are freezing, tiring or starving.

Sacerdotal celibacy in the Christian Church and the monastic tradition generally are the most obvious manifestation of this war against the libido.[6,7] I have contended elsewhere that the roots of sacerdotal celibacy in Christianity are plainly Gnostic: the clearest instance of the dangerously incomplete victory over the Gnostics in the first few centuries of the Church's life.[8]

Inevitably, the more intense the suppression of natural sexuality, the more intense the religious feeling that comes with the rechannelling of libido towards God. St Teresa and St John of the Cross, the most notorious Christian eroticists, were zealous reformers of the Carmelites: they thought the Order, as it was in the sixteenth century, was too soft. They insisted, for instance, on going barefoot, calling themselves the 'discalced' (barefoot) Carmelites to distinguish themselves from the calced Carmelites whose contemptible surrender to the flesh went to the extent of wearing shoes in winter.

There's abstinence and abstinence. A particularly perverse and accordingly allegedly powerful means of redirecting sexual energy was to have intercourse but to stop short of orgasm.

But all this is extreme stuff. The vast majority of faithful religious people over the millennia have seen sex as a God-given cup from which to drink: it is simply raised to him in toast. For a few more poetic ones there has been the justification of Diotima, whose argument is voiced by Socrates in the *Symposium*. Sex can be a route to a sort of elevated, intellectual apprehension of the Divine. Approach the capricious god Eros in the right way, she said, and he can help you to:

> ascend from the things of this world until you begin to catch sight of that beauty, and then you're almost within striking distance of the goal. The proper way to go about or be guided through the ways of love is to start with beautiful things in this world and always make the beauty I've been talking about the reason for your ascent. You start by loving one attractive body and step up to two; from there you move on to the beauty of people's activities, from there to the beauty of intellectual endeavours, and from there you ascend to that final intellectual endeavour, which is no more and no less than the study of that beauty, so that you finally recognise true beauty.[9]

There's not much of Plato I can read, these days, without sprinkling it first with so many caveats that it doesn't taste much like Plato by the time I tuck in. And this is no exception. It's the sort of thing that could only be written by a sanctimonious bore who didn't like people much and didn't know how to enjoy either himself or anything else. It's a crying, ironic shame that Christianity, which of all creeds should be the most hostile to

Plato, has adopted and hallowed him.[10] When the Forms creep into the bedroom, Eros, Dionysus and the lovers of the Song of Songs bolt.

Most orthodox Judaism will have no truck with Plato, although of course the Hasidim have had more than a flirtation with him. In most Jewish heads, today is the day of salvation, because there is no other day. And in Christianity, although there may be other days, today is the day of salvation too. In both Jewish and Christian bedrooms, tonight is the night of consummation, because the night is good in itself. You needn't talk too earnestly about the transports of erotic delight. The night needn't take you anywhere except closer to your lover. Your lover is not God, but God won't mind too much if, at least for a moment, you forget it. Jewish and Christian lovers laugh affectionately at Monica, really because she takes herself rather too seriously. They know too that if you're looking for what Monica is looking for as hard as she is, you're likely to miss it altogether. There are some truths so elusive that they can only be grasped when you're laughing in bed. They'd get her to throw away her briefcase and fold her Post-It notes into flowers.

CHAPTER 8

Finding God in the Intensive Care Unit: Near-death and Other Out-of-body Experiences

The blackness began to erupt into a myriad of stars and I felt as if I were at the center of the universe with a complete panoramic view in all directions. The next instant I began to feel a forward surge of movement. The stars seemed to fly past me so rapidly that they formed a tunnel around me. I began to sense awareness, knowledge. The further forward I was propelled the more knowledge I received. My mind felt like a sponge, growing and expanding in size with each addition. The knowledge came in single words and in whole idea blocks. I just seemed to be able to understand everything as it was being soaked up or absorbed. I could feel my mind expanding and absorbing and each new piece of information somehow seemed to belong. It was as if I had known already but forgotten or mislaid it, as if it were waiting here for me to pick it up on my way by.

Virginia Rivers[1]

Sometime in the first century, from Macedonia or Ephesus, St Paul wrote to the Corinthian Christians:

I know a person in Christ who fourteen years ago was caught up to the third heaven – whether in the body or out of the body I do not know; God knows. And I know that such a person – whether in the body or out of the body I do not know; God knows – was caught up into Paradise and heard things that are not to be told, that no mortal is permitted to repeat.[2]

He is talking about himself.

In the early twenty-first century, in a hospital in London, my wife, in labour with our second child, and inhaling the mixture of nitrous oxide and air given for analgesia, felt 'herself' rising out of her body. She looked down at her body with detached clinical interest (she's a doctor), and she swears that she saw our son's head appearing for the first time. If she was in her body, that would have been impossible.

'Out-of-body experiences' (OBEs) are common.[3] About 10 per cent of us will have such an experience at some stage in our life.[4] Icelanders have the fewest; the users of epic quantities of cannabis have the most. They often occur in the strange no-man's-land between sleep and waking – an association exploited by several visionaries. Thomas Edison, seeking inspiration, would doze off with a dollar coin balanced on his head and a metal bucket on his lap. As his head lolled forward, the coin would fall into the bucket, the clang would summon him back from sleep, and in the moment of not-quite-awake he would often burst through the blockages to a new solution. Salvador Dali used to sleep in a chair with his chin resting on a spoon held in his hand. When he went to sleep his chin lurched forward, propelling him into a hypnagogic world populated by his ghosts and chimaeras. He snatched a few strands of dream, sprang up, and began to paint.[5]

There are other artificial means of induction. By arduous occult training it is possible to haul oneself out of one's body at will: some can climb a rope out of themselves, leaving their bodies on the couch; some can prise their bodies apart from their 'soul', letting the untethered soul drift to the ceiling; some can project their 'souls' like lasers. Sensory deprivation or sensory hypersaturation might help. Absolute silence, white noise, flotation tanks[6] or the sort of mummification used in some forms of sexual bondage practice can gently loosen the adhesive binding body and 'soul' together; extreme physical effort and pain can wrench them apart. A doctor of impeccable respectability told me that during the seduction scene in *Tristan and Isolde* in Covent Garden, he left his body sitting in the most expensive seat in the stalls and watched the scene, and the top of his own head, from several feet above. He put it down to profound erotic and emotional arousal. His body just wasn't an adequate vessel for the passion that he felt, and so it was temporarily discarded.

Dissociative agents such as ketamine work very reliably as dividers of body and 'self' – so much so that when used in human anaesthesia, amnesiac drugs are given too to wipe the sometimes alarming memory of hovering above one's own body on the operating table.

Scientific work on OBEs has been patchy and speculative. There is little hard data. It has been claimed that lesions and consequent epileptiform activity at the junction of the right parietal and temporal lobes can induce OBEs.[7] The junction seems to be the place, or a place, where information about the position of the 'self' and the position of the body coming in from the various sensors is brought together and processed. If there is a mismatch between the co-ordinates for the 'self' and the co-ordinates for the body, an OBE may result. This chimes well enough with Newberg's claim that reduced parietal lobe activity might blur the frontiers of the self.[8]

A promising line of inquiry comes from the experimental induction of OBEs in healthy subjects.

Imagine yourself sitting in a laboratory. Over each of your eyes is a small screen. Onto those screens is projected a live film of your own back taken by two video cameras placed next to one another two metres behind your head. The left-hand camera projects to the left-eye screen, the right-hand camera to the right. You see these images as a single three-dimensional image of yourself. Imagine that a researcher then stands to your side.

You can see him. He has a rod in each hand. Simultaneously he taps your chest (although you have no direct view of him doing so) and moves the other rod towards the place (off camera) where the chest of the image would be. You will feel that you are sitting outside your physical body, viewing it from the perspective of the cameras.[9]

While interesting, it is not clear how significant this is. The context is a million miles from sleep, bondage, ketamine or the deathbed. Which brings us to near-death experiences (NDEs).

They are tremendously common. Studies have suggested that around 50 per cent of those who have come very near death and are asked nicely will report an NDE.[10] A study in *The Lancet* summarises the circumstances in which NDEs, and related experiences arise:

> NDE[s] are reported in many circumstances: cardiac arrest in myocardial infarction (clinical death), shock in postpartum loss of blood or in perioperative complications, septic or anaphylactic shock, electrocution, coma resulting from traumatic brain damage, intracerebral haemorrhage or cerebral infarction, attempted suicide, near-drowning or asphyxia, and apnoea. Such experiences are also reported by patients with serious but not immediately life-threatening diseases, in those with serious depression, or without clear cause in fully conscious people. Similar experiences to near-death ones can occur during the terminal phase of illness, and are called death-bed visions. Identical experiences to NDE[s], so-called fear-death experiences, are mainly reported after situations in which death seemed inevitable: serious traffic accidents, mountaineering accidents, or isolation such as with shipwreck.[11]

The logical positivist philosopher A. J. Ayer spent most of his life haughtily dismissing religion. Shortly before his actual death he had an NDE. 'My recent experiences', he wrote, 'have slightly weakened my conviction that my genuine death, which is due fairly soon, will be the end of me, though I continue to hope that it will be. They have not weakened my conviction that there is no god.'[12]

This seems clear enough, but Ayer was evidently embarrassed by it. Perhaps he hated the idea of being recruited by Dawkins' 'faith heads'. In any event, a few days later, he hastily corrected himself: '[W]hat I should have said is that my experiences have weakened, not my belief that there is no life after death, but my inflexible attitude towards that belief.' A nice distinction indeed. Whatever had happened, it had plainly shaken him to the core. His attending physician, Dr Jeremy George, suggested that the effect had been far greater than Ayer's own measured prose indicated. Ayer had told him, said George, 'I saw a divine being. I'm afraid I'm going to have to revise all my books and opinions.'

Probably the oldest explicit account of an NDE that we have is in Plato's *Republic*.[13] A soldier, Er, is slain in battle. Ten days after his 'death', when the decomposed bodies of his comrades in arms have already been removed, his corpse is found, unaffected by decay, and carried off home for the funeral. On the twelfth day, during his own funeral, he returns to life, telling the astonished mourners that he had been to the other world. Socrates takes up the story.

> He said that when his soul left the body he went on a journey with a great company, and that they came to a mysterious place at which there were two openings in the earth; they were near together, and over against them were two other openings in the heaven above. In the intermediate space there were judges seated, who commanded the just, after they had given judgment on them and had bound their sentences in front of them, to ascend by the heavenly way on the right hand; and in like manner the unjust were bidden by them to descend by the lower way on the

left hand; these also bore the symbols of their deeds, but fastened
on their backs. He drew near, and they told him that he was to
be the messenger who would carry the report of the other world
to men, and they bade him hear and see all that was to be heard
and seen in that place. Then he beheld and saw on one side the
souls departing at either opening of heaven and earth when
sentence had been given on them; and at the two other open-
ings other souls, some ascending out of the earth dusty and
worn with travel, some descending out of heaven clean and
bright. And arriving ever and anon they seemed to have come
from a long journey, and they went forth with gladness into the
meadow, where they encamped as at a festival; and those who
knew one another embraced and conversed, the souls which
came from earth curiously enquiring about the things above,
and the souls which came from heaven about the things beneath.
And they told one another of what had happened by the way,
those from below weeping and sorrowing at the remembrance
of the things which they had endured and seen in their journey
beneath the earth (now the journey lasted a thousand years),
while those from above were describing heavenly delights and
visions of inconceivable beauty.[14]

Today, NDEs are very commonly reported following cardiac
arrest and resuscitation. Both the scientific and the lay literature are full of stories. Two examples make the point.

The Lancet quoted a nurse's story. A forty-four-year-old man
was brought to hospital by ambulance, having been found about
an hour before, lying in a field. He was comatose when admitted.
Heart massage was begun, and he was defibrillated. An endo-
tracheal tube was then inserted. In order to do so the patient's
dentures were removed and put onto the crash trolley while
aggressive CPR was continued. After about an hour and a half
of this, the patient had reassuring heart rhythm and blood pressure, but was still intubated, ventilated and comatose. He was

transferred to the intensive care unit for continued ventilation. It was more than a week later that the nurse who had removed the dentures saw the patient again.

> 'The moment he sees me,' said the nurse, 'he says: "Oh, that nurse knows where my dentures are." I am very surprised. Then he elucidates: "Yes, you were there when I was brought into hospital and you took my dentures out of my mouth and put them onto that car. It had all these bottles on it and there was this sliding drawer underneath and there you put my teeth." I was especially amazed because I remembered this happening while the man was in deep coma and in the process of CPR. When I asked further, it appeared the man had seen himself lying in bed, that he had perceived from above how nurses and doctors had been busy with CPR. He was also able to describe correctly and in detail the small room in which he had been resuscitated as well as the appearance of those present like myself. At the time that he observed the situation he had been very much afraid that we would stop CPR and that he would die. And it is true that we had been very negative about the patient's prognosis due to his very poor medical condition when admitted. The patient tells me that he desperately and unsuccessfully tried to make it clear to us that he was still alive and that we should continue CPR. He is deeply impressed by his experience and says he is no longer afraid of death. Four weeks later he left hospital as a healthy man.'[15]

One of the most spectacular NDE accounts is that of Pam Reynolds, who had surgery to remove a huge aneurysm thumping away in a very inaccessible part of her brain.[16] The technique used involved lowering her body temperature to 15 degrees centigrade, and stopping her heart and her breathing. During the procedure her electroencephalogram was flat. Her brain, in other words, showed no electrical signs of life. A surgical saw was used to cut through her skull, the aneurysm was duly excised and Reynolds was resuscitated. When she was able to speak, she told a remarkable story.

The electrical saw, she said, made a perfect 'D' note, and as she listened to the note, it seemed to be pulling her out of the top of her own head. The tone got clearer the further out of her body she came. 'I had the impression [that the note] was like a road, a frequency that you go on.' She hovered over the operating table at the height of the surgeon's shoulder, watching what was going on. She had a perfect view; in fact more than a perfect view: 'It was not like normal vision. It was brighter and more focused and clearer than normal vision.' She looked with interest at all the medical busyness, commenting ruefully to herself that they had shaved her head in a very odd way. She noted and later remembered minute details of it, the electrical-toothbrush-like saw. She heard comments about the size of her blood vessels, saw the cardiac bypass machine and the ventilator, and then the next stage of her journey began.

She felt a pulling sensation, but there was no coercion. She went along with it because she wanted to go. She later groped for similes. 'It was like the Wizard of Oz – being taken up in a tornado vortex, only you're not spinning round like you've got vertigo. You're very focused and you have a place to go.' It was like going up very fast in a lift. 'And there was a sensation, but it wasn't a bodily, physical sensation. It was like a tunnel, but it wasn't a tunnel.'

Near the start of the tunnel vortex, Pam became aware that her grandmother was calling her. But she didn't hear it with her ears. It was clearer than hearing: 'I trust that sense more than I trust my own ears.' Pam felt that her grandmother wanted her to come to her, so she continued along the dark shaft, entirely without fear. At the far end there was a small bead of bright light that got bigger as she approached it. The light became

An artist's representation of a near-death experience: a friend or relative waits in the light at the end of the tunnel to greet the traveller.

intense, 'like sitting in the middle of a light bulb'. Slowly she began to see figures in the light: 'they were all covered with light, they *were* light, and had light permeating all around them'. They took shapes. One of them was her grandmother, but there were others too: a grandfather, an uncle, a cousin. They were looking after her. Each looked as they had looked at the best time of their life. They wouldn't let Pam go any further. They didn't speak to her, but indicated clearly in some other way that if she went all the way into the light, 'something would happen to me physically. They would be unable to put this "me" back into the body "me", like I had gone too far and they couldn't reconnect.'

Pam wanted to go on into the light, but also recognised that she had a duty to return – particularly to care for her children. Her dead relatives then fed her – but not through her mouth –

with something she described as 'sparkly'. She was being nour-
ished somehow, and made strong. She felt ready for anything.

She expected her grandmother to guide her back through the
tunnel, but for some reason her grandmother 'just didn't think
she would do that'. Her uncle said he would go instead, and
they went back together. 'But then I got to the end of [the
tunnel] and saw the thing – my body. I didn't want to get into
it . . . It looked terrible, like a train wreck. It looked like what
it was – dead.' It was covered up, and Pam was scared of it and
didn't want to look at it.

It was 'communicated' to her that getting back into her body
was like jumping into a swimming pool. She was reluctant, and
eventually had to be pushed by her uncle. 'The body was pulling
and the tunnel was pushing . . . It was like diving into a pool
of ice water . . . It hurt!' In the operating theatre, a tape of
'Hotel California' was blaring away.

Pam Reynolds' account, although dramatic, follows the general
pattern seen in NDEs, and indeed in many OBEs, however
induced. Kenneth Ring identified five broad stages.[17] Often the
experience begins with feelings of contentment and peace. There
is then some sort of detachment from the body. The subject
may feel their 'self' being pulled slowly out of their body, or
explosively shot out, and will often look down on their own
body. Then comes the famous 'tunnel', which we have met already
and will meet again when looking at shamanic journeys,
migraine, epilepsy and the action of psychedelic drugs. The
tunnel is typically dark and spiralling. It may have cross-
hatchings on the side. The subject travels fast towards the end,
where there is a bright light. Finally, the subject may float or
walk into the light.

Not all near-death-experiencers go through all stages: 60 per
cent get to the first stage; about 10 per cent get to the last.[18]
There are many variations on the basic pattern. Subjects in the

tunnel are often reassured by the presence of spiritual figures, dead friends or family members. They may meet a Being, often dazzlingly white, and often perceived by them as whatever deity they believe in. Their lives might be replayed before them, with a sensation of judgement. The whole experience is usually comfortable, and subjects generally return only reluctantly to their own body – which may feel cold and unwelcoming. Often subjects feel that they have been returned to their own body for a purpose. Christians are often told that their time has not yet come; Indians, apparently, that there has been a clerical error.[19]

These are the accounts of NDEs that you get in the popular books on the subject – usually with soft-focus flowers on the cover. But it is not always that way.[20] The NDEs of the Middle Ages were often full of horrors: tortures, cackling demons and a crushing sense of separation.[21] Modern NDEs resulting from suicide attempts are sometimes associated with terrors. 'If you leave here a tormented soul you will be a tormented soul over there too,' one woman told Raymond Moody.[22]

The content of an NDE depends to some extent on the person experiencing it. If a Hindu sees a deity, it is likely to be one from the Hindu pantheon rather than the figure of Jesus. Children do get NDEs, but they are less complex and less laden with spiritual overtones than those of adults. Children are more likely to talk to their (still living) siblings or parents than they are to talk to Jesus. It has been noted that modern children are less likely than children a few decades ago to meet dead people in their NDEs – presumably because modern children are unlikely to have had friends who have died.[23] It has been reported that a high proportion of blind people having NDEs have some sort of visual awareness[24] – an observation predictably seized upon by those wishing to establish that consciousness survives death, and equally predictably savaged by those who wonder loudly how meaningful such reports can be. Blind people aren't blind in their dreams, they say, so why is it significant that they say they saw when they were unconscious on an operating table?

One very common report is that the NDE has produced in the subject a profound life change. They often become more religious in a conventional sense. 'Most near-death survivors say that they don't think there is a God,' said Nancy Evans Bush, of the International Association for Near-Death Studies. 'They know.'[25] Subjects typically say that they are less selfish, less materialistic, more loving, more giving. It is not always good for them. People who have had NDEs 'can see the good in all people', observed psychiatrist and noted NDE researcher Bruce Greyson. 'They act fairly naive, and they often allow themselves to be opened up to con men who abuse their trust.' Divorce rates are higher, and there are often workplace problems. It is not that the subjects are stupid or negligent, says Greyson; it is that their values have been radically transformed. 'The values you get from an NDE are not the ones you need to function in everyday life.'[26] If you think that you have climbed Jacob's ladder, it is hard to think that the next rung on the promotion ladder matters much.

It has been suggested that some categories of people are more likely to get NDEs than others. A 2006 study, reported in *Neurology*, compared fifty-five people who had had NDEs with an equal number who had not. The findings squared very elegantly with the work on OBEs and sleep. (Remember Edison and Dali?) Those who had had NDEs were much more likely than the others to have blurred boundaries between sleeping and waking.[27] But being religious or of any particular religion won't improve or diminish your chance of having an NDE. Nor will being of any particular gender or ethnic group.[28]

But having an NDE is not the same as being able to report it. The experience, obviously, has to be remembered in order to be reported. And that may be the explanation for a finding reported in the respected journal *The Lancet*, to the effect that the frequency of reported NDE was higher in younger than in older survivors of cardiac arrest. Perhaps everyone brought back from the brink by enthusiastic resuscitation had an NDE, but only the younger ones had sufficient short-term memory to recall and relate it.[29]

Early scientific work on the near-death scenario focused on deathbed visions – reports that the dying, typically lying on the deathbed at home, surrounded by their families, saw vistas of the other world and conversed with dead friends and relatives. This work, bar some quaint attempts to weigh the soul, was largely anecdotal.[30] Some bystanders said that they saw the spirit leaving the body and heard angelic music, often of a clichéd tinkling or trumpeting kind, as the patient crossed to the other side. Deathbed visions, at least in the West, seem to have been casualties of modern medicine. Thanks to the available armoury of palliative drugs, and the sheer ability to keep people alive longer, far fewer people are fully conscious immediately before their deaths.[31] The demise of the deathbed vision, though, has coincided with many more reports of the classic NDE.

One question that has repeatedly been asked but not definitively answered is whether the 'self' that seems to leave the body and hover above it can actually perceive things that it could not have guessed, subsequently reconstructed or otherwise gleaned. There are plenty of tales (Pam Reynolds is an excellent example) about 'clinically dead' patients reporting events of which, according to their flat EEGs, they should have been ignorant. But these are difficult to evaluate. Newspapers and TV screens are full of information about medical and surgical practices. Most people will, consciously or unconsciously, have absorbed some of it. Imagination is terrifyingly powerful. Its ability to convince the imaginer is awesome. We cling to no delusions as strongly as the ones we have concocted ourselves.

And we really don't know how sensitive a flat EEG is as an index of our ability to receive information. All that a flat EEG really tells us is that there is a flat EEG.

When brain perfusion drops, our ability to perceive things apparently diminishes too, but one of the last senses to go is our hearing. Possibly some of the NDEs in which surgical tools were apparently seen were imaginative visual reconstructions from data funnelled entirely through the auditory pathways. Surgeons and nurses chat merrily away all the time: the vibration in your skull from an electric saw buzzing into it can only be produced in a certain very limited number of ways. In ordinary living, if we hear a sound produced by something out of sight, our brain conjures a visual image of whatever has produced the sound. It has not been demonstrated that a similar process cannot occur in the hypoxic brain.

What would be impressive and paradigm-shattering would be evidence that the disembodied self had seen (for instance) a five-digit number on an otherwise inaccessible shelf above which the self is said to have hovered. That is precisely what is said to have happened in an OBE in a young woman, prone to OBEs, being examined in a sleep laboratory in California.[32] Although the work has not been disproved, it has been suggested that the subject could somehow have cheated. We might soon have some better data, however, since, in an unprecedentedly enlightened burst of enthusiasm for paranormal research in England, a five-year study was launched in 2008 which will look at near-death experiences in patients who have had a cardiac arrest.[33] It will include an evaluation of whether the disembodied 'soul' can see things that the body cannot.

There has been no shortage of scientific and lay speculation about the origin of NDEs. Some have postulated that the experience is the final biochemical wave of the dying brain – perhaps mediated by compounds released by brain cells on their last metabolic legs.[34] Others think that the phenomenon is mainly psychological – a desperate attempt by whatever the self is to wrestle with entirely unprecedented challenges to its very existence.[35] Perhaps the self is so constitutionally incapable of contemplating its own annihilation that, faced with a diminution in its basic elements – a dilution of self by the invading tide of eternity – it goes into denial, telling itself a comforting fairy tale of survival.

The whole experience has been poetically seen as almost literal rebirth – a thesis that has rhetorical force in the minds of some because it resonates conveniently with notions of reincarnation and the Christian idea that one must be reborn to enter the kingdom of heaven. In an NDE, the theory goes, you are reliving the moment of your biological birth. The tunnel is your mother's vagina, the light at the end is the light of the labour suite, and the various people you meet are personifications of the kindly midwife and any others clustering round at the business end.[36] Unfortunately it is nonsense. During birth the baby's eyes are closed, but even if they were not, they would not be peering down a tunnel. The side of the birth canal is generally applied closely and claustrophobically to the face, and it is the top of the head, not the face, that first dives out into the light. Susan Blackmore gleefully drove the near-final nail into the coffin of this theory when reporting her survey showing that, of people who had had OBEs with tunnel experiences, the proportion who had been born by Caesarean section was almost exactly the same as the proportion who had had a vaginal birth.[37]

The most detailed and physiologically coherent attempt to reconstruct and explain NDEs has probably come from Susan Blackmore.

First the tunnel. We start with the rather trite observation that images from the outside world map onto the retina, and the retinal image maps in turn onto the brain's visual cortex.

In a normal, active brain the relationship between these sets of images is kept pretty constant. But when the brain is deprived of blood, and therefore oxygen, the normal feedback systems which stop the brain from running completely wild begin to malfunction.[38] It takes metabolically expensive policemen to stop the brain from turning anarchic. If you can't pay the police force in hard oxygen, the streets erupt. Since we know, more or less, the mathematics of the world-retina-cortex mapping, we can predict that neurological anarchy will result in stripes of activity moving across the cortex. How will those stripes be interpreted by the higher processing centres? We know that too: the brain will conclude that the eyes are looking at a spiralling tunnel.[39]

The British psychologist Susan Blackmore, who proposes an entirely natural explanation for near-death and other out-of-body experiences.

Susan Blackmore and Tom Troscianko built on this model. They devised a computer program that simulated what would happen if you had gradually increasing electrical anarchy in the visual cortex, but, crucially, built in the anatomical fact that the centre of the visual field is represented by many cells, but the edge of the field by very few. This is Blackmore's description:

> The computer program starts with thinly spread dots of light, mapped in the same way as the cortex . . . Gradually the number of dots increases, mimicking the increasing [electrical] noise. Now the centre begins to look like a white blob and the outer edges gradually get more and more dots. And so it expands until eventually the whole screen is filled with light. The appearance is just like a dark speckly tunnel with a white light at the end, and the light grows bigger and bigger (or nearer and nearer) until it fills the whole screen.[40]

This model, to me, has great elegance and explanatory power. And of course, since tunnels are so ubiquitous in so-called mystical experience, the explanation is repercussive a long way from the coronary care unit.

Next, the out-of-body experience: the hovering over oneself. We think we know what the world 'out there' is like. But we don't. We have already seen that what we 'know' is the result of our brain's processing of the data beamed into it by our senses.[41] I think I know that I am sitting in an ancient library. But in fact all I know about where I am is that my brain has assembled a model which has me sitting here: it is not knowledge about where I am at all. But, to stave off madness, I label my model 'real'. I choose to apply the label to this model instead of any other because it is more stable than the others. It's more solid and reliable, although perhaps rather more dull than other contenders. I know this model, and I know that it works, after a fashion.

Normally, while my brain is properly perfused and not drenched with psychedelic drugs, I can distinguish between this 'real' model and the other things that come flitting through my consciousness. A moment ago I had a picture of myself swimming with humpback whales in the Quirimbas Archipelago in Mozambique. But I had no difficulty in acknowledging that, sadly, I wasn't really there. That picture was of a different quality from the model of me slumped in the bookstacks. All the time different models of reality compete for prominence in my head, but at least until the pubs open or I have a myocardial infarction, they are unlikely to displace the library model (which also encompasses Radcliffe Square, the cycle ride home, my study at the top of the house, screaming children, and so on). The library model – the 'real' model – is much more stable than its competitors; sometimes infuriatingly so.

But it will not always be as stable. Its sovereignty will be seriously threatened by anoxia, the by-products of anaerobic metabolism and the whole pharmacopoeia of fear and stress. Then all the models will be unstable. So what happens then? If I am dying, and the blood supply to my brain is failing, a tunnel will be created by electrical noise in my visual cortex. Blackmore suggests that the tunnel will be the most stable model, and that I will accordingly perceive it as 'real'. It will have the same quality for me then as the library model has for me now.

That constantly bubbling, questioning brain – the brain so dangerously unable to cope with uncertainty – will want to know what is happening to it. It will try to understand. And to do that it will fumble through its memory banks. Probably it will be able to recover fragments of what happened: the chest pain; the last swerve of the car before the collision; the frantic attempts to defibrillate me. On the skeleton of these fragments it will try to build an explanation. Susan Blackmore says:

We know something very interesting about memory models. Often they are constructed in a bird's eye view. That is, the events or scenes are seen as though from above. If you find this strange, try to remember the last time you went to a pub or the last time you walked along the seashore. Where are 'you' looking from in this recalled scene? If you are looking from above you will see what I mean.

So my explanation of the OBE becomes clear. A memory model in bird's eye view has taken over from the sensory model. It seems perfectly real because it is the best model the system has got at the time. Indeed, it seems real for just the same reason anything ever seems real.[42]

As for the 'life review', perhaps the rioting, dying brain stirs up memories in dusty, long-forgotten archives. Temporal lobe excitation is known to generate sensations that seem like (and perhaps are) the replaying of old memory tapes. Kinseher has recently elaborated on this, contending that the alarmed brain, confronting the terrifying novelty of dying, embarks on a frantic trawl of the memory in the hope of coming up with a comparable experience which it can use as a benchmark to assess the experience.[43]

As for the feelings of bliss, they are perhaps the easiest of all to explain. A stressed brain is likely to gush with precisely the sorts of neurotransmitters that are known to mediate pleasurable sensations. And the brain may be bathed in endogenous opiates too – giving a sort of terminal heroin high.[44]

And finally, ineffability. 'What I met then', comes a repeated refrain from the far frontiers of life, 'is too wonderful to put into words.' I hesitate to follow Susan Blackmore into this part of her reconstruction (as appears in the Appendix to this book, on the problem of consciousness, I would insert many caveats), but as a corollary of what we know about the nature of consciousness, the basic thesis is compelling.

All the mental models in any person's mind are conscious, but only one is a model of 'me'. This is the one that I think of as myself and to which I relate everything else. It gives a core to my life. It allows me to think that I am a person, something that lives on all the time. It allows me to ignore the fact that 'I' change from moment to moment and even disappear every night in sleep. Now when the brain comes close to death, this model of self may simply fall apart. Now there is no self. It is a strange and dramatic experience. For there is no longer an experience – yet there is experience. This state is obviously hard to describe, for the 'you' who is trying to describe it cannot imagine not being.[45]

So far so good. But there is a danger in thinking that, by coming up with a plausible hypothesis for each of the common elements of an NDE, the oddness drains out of the whole phenomenon. It is not so. Even if we remind ourselves that neurological disinhibition is by definition likely to generate some counter-intuitive effects, there are still some profound mysteries here. Greyson summarises well the position adopted by many:

No one physiological or psychological model by itself explains all the common features of NDE. The paradoxical occurrence of heightened, lucid awareness and logical thought processes during a period of impaired cerebral perfusion raises perplexing questions for our current understanding of consciousness and its relation to brain function. A clear sensorium and complex perceptual processes during a period of apparent clinical death challenge the concept that consciousness is localized exclusively in the brain.[46]

I would like to agree with him. Indeed, I agree with his bottom line – that consciousness is not located in the brain. But NDEs are not authority for that proposition. The authority of NDEs as spiritual teachers has been dealt a serious blow by a commonly used veterinary anaesthetic – ketamine. Ketamine is deeply embarrassing to the NDE industry, and the ketamine papers have by and large been swept quickly under the carpet.

Some of ketamine's effects have already been described.[47] It works as an anaesthetic partly because patients become so 'dissociated', or removed from 'their' body, that they don't object to the surgeons hacking it up.[48] It is therefore wholly different in action from consciousness-ablating anaesthetics.

The point is that the effects of ketamine seem to be very similar indeed to NDEs. The psychiatrist Karl Jansen notes:

> Ketamine administered by intravenous injection is capable of reproducing all of the features of the NDE which have been commonly described . . . The present author has experienced several NDEs and has also been administered ketamine as an anaesthetic and within experimental paradigms. The NDEs and ketamine experiences were clearly the same type of altered state of consciousness. Ketamine repeatedly produced effects which were like the NDEs described by [all the classic collectors of information about NDEs, whose work has been summarised above]. Ketamine reproduced travel through a tunnel (sometimes described as 'the plumbing of the world', or in mundane terms such as 'like being on a subway train'), emergence into the light, and a 'telepathic' exchange with an entity which could be described as 'God'.[49]

Perhaps equally significantly, in the context of our inquiry into the biology of spiritual experience generally, Jansen comments:

> Neither the NDEs nor the ketamine experiences bore any resemblance to the effects of psychedelic drugs such as dimethyltryptamine (DMT; also administered to the author in experimental paradigms), and lysergic acid diethylamide (LSD).[50]

Jansen is one of very many voices making the comparison. The comparison is overwhelming. There is apparently complete overlap. But what does the overlap mean? It suggests very strongly that the neurophysiology of NDE is the same as that of a ketamine trip: the two experiences use broadly the same routes.

The major binding site of ketamine in the brain is the phencyclidine site on the NMDA receptor. The role of NMDA receptors is slowly being unravelled, but this book is no place for an essay on them. It is enough to say that they are known to be important in psychoses, epilepsy, and in the cell death that results from anoxia and epilepsy. It all seems to fit.

We ended Chapter 6 by noting that if both X and Y use the same road, it does not mean that X and Y are the same thing. That caution is appropriate here too. But surely the similarity here is more than mere use of the same street. The car is identical; the registration number is the same; the drivers are indistinguishable to the naked eye.

Unless one is to say that God lives in the fridge of your local veterinary surgeon, packaged up in 10 ml vials, you have a real problem ascribing world-shattering religious significance to NDEs.

CHAPTER 9

Other Portals

The Professor, who was a very remarkable man, didn't tell them not to be silly or not to tell lies, but believed the whole story. 'No,' he said, 'I don't think it will be any good trying to go back through the wardrobe door to get the coats. You won't get into Narnia again by *that* route ... Yes, of course you'll get back to Narnia again some day...'

(C. S. Lewis, *The Lion, the Witch and the Wardrobe*, 1950)

On 2 April 1968, the Virgin Mary was noticed to be hovering over the Coptic Orthodox church of St Mary in Zeitoun, a district of Cairo. She continued to appear there intermittently for the next couple of years – sometimes for a few minutes, and sometimes for several hours – sometimes attended by lights which, to the eyes of the very numerous faithful, seemed to be doves. She was seen by millions – and not just Copts or Catholics. The blind saw too, and the lame ran to tell their stories: there were many reports of healings. Many turned to Christianity. Sophisticated investigations by apparently independent investigators came up with no obvious cause.

In 1917, the Virgin Mary allegedly appeared to three shepherd children in Fatima, Portugal. She was 'brighter than the sun', said one of the children, 'shedding rays of light clearer and stronger than a crystal ball filled with the most sparkling water and pierced by the burning rays of the sun'. She explained to the children that they would be the bridgehead for a great revival, and exhorted them to various penances to prepare the way, such as wearing painfully constricting bands around their wrists and abstaining from water on hot days. (Pain and dehydration are, of course, known to be potent triggers of altered states of consciousness.) The appearances happened every thirteenth day of the month for six months, and then the Virgin promised a spectacular apparition that would turn many to faith.

The three children who allegedly saw the Virgin Mary in Fatima, Portugal, in 1917.

Word got around. A crowd of about seventy thousand gathered in the countryside on 13 October 1917. 'Look at the sun!' cried one of the children, and pointed. A reporter for one of the Lisbon dailies takes up the story:

[T]he silver sun, enveloped in the same gauzy grey light, was seen to whirl and turn in the circle of broken clouds ... The light turned a beautiful blue, as if it had come through the stained-glass windows of a cathedral, and spread itself over the people who knelt with outstretched hands ... people wept and prayed with uncovered heads, in the presence of a miracle they had awaited. The seconds seemed like hours, so vivid were they.[1]

Some said that the sun fell to earth; others that it zigzagged away out of sight.

Now those, says Michael Persinger (he of the God helmet and the three-piece suit), are classic examples of apparitions generated by tectonic strain. Near seismic faults, he contends, the earth is under tremendous strain, and one of the results is a very strong electromagnetic field. These fields can themselves produce lights of the sort seen at Zeitoun and Fatima, and can also play hallucinogenic games with the temporal lobes of all the watchers, so that they all 'see' lights at the same time. Many UFO sightings, he says, can be explained in this way.

Tectonic strain theory has had an icy reception from scientific critics. There are two broad strands of criticism. First, to claim a particular phenomenon as a creature of tectonic strain, Persinger has to say that you can see UFO-type lights hundreds of miles away from the relevant area of seismic busyness. But since there are few, if any, places on earth more than a few hundred miles away from such an area, the alleged correlation between seismic activity and mysterious lights evaporates. Persinger has no very robust riposte. And second, we are surrounded by much more powerful electromagnetic fields than any that could conceivably be generated by a distant surge of molten metal in the earth's crust. If Persinger is right, why don't you see flying saucers whenever you switch on your cell phone or go to sleep with your electric watch next to your head?[2]

Persinger's response is simply to say that size isn't everything, and that the nature of the field might matter more than its magnitude. To that his critics have had no very robust riposte, but you can't blame them. It's a slippery idea, likely to evade definitive disproof.

Apparitions of the sort recorded at Zeitoun have played a central part in human religious history. We have looked at several biblical examples, but there are plenty of others. Here is one.

On 21 September 1823, Joseph Smith was praying. He asked God specifically for a divine manifestation, of the sort he had experienced before. Smith needed encouragement – others were finding his earlier stories of a meeting with God difficult to take – and he was confident that God would not let him down. Expectation is known to help hallucination, and whether or not it contributed here, the manifestation was spectacular.

> While I was thus in the act of calling upon God, I discovered a light appearing in my room, which continued to increase until the room was lighter than at noonday, when immediately a personage appeared at my bedside, standing in the air, for his feet did not touch the floor.
>
> He had on a loose robe of most exquisite whiteness. It was a whiteness beyond anything earthly I had ever seen; nor do I believe that any earthly thing could be made to appear so exceedingly white and brilliant. His hands were naked, and his arms also, a little above the wrist; so, also, were his feet naked, as were his legs, a little above the ankles. His head and neck were also bare. I could discover that he had no other clothing on but this robe, as it was open, so that I could see into his bosom.
>
> Not only was his robe exceedingly white, but his whole person was glorious beyond description, and his countenance truly like lightning. The room was exceedingly light, but not so very

bright as immediately around his person. When I first looked upon him, I was afraid; but the fear soon left me.

He called me by name, and said unto me that he was a messenger sent from the presence of God to me, and that his name was Moroni; that God had a work for me to do; and that my name should be had for good and evil among all nations, kindreds, and tongues, or that it should be both good and evil spoken of among all people.

He said there was a book deposited, written upon gold plates, giving an account of the former inhabitants of this continent, and the source from whence they sprang. He also said that the fulness of the everlasting Gospel was contained in it, as delivered by the Savior to the ancient inhabitants. . .[3]

And so the Mormon religion was born.

There's an unfortunate modern tendency, faced with stories like this, to shout, 'Pathology!' It's pejorative and medically illiterate. Quite apart from discounting irrationally the possibility that the apparition was objectively present, it forgets that hallucinations are surprisingly common in apparently sane people. If hearing voices means that you are schizophrenic, for instance, many of us are. Bentall and Slade noted that over 15 per cent of male students agreed with the statement that, 'In the past I have had the experience of hearing a person's voice and then found that no one was there'. Nearly 18 per cent agreed with, 'I often hear a voice speaking my thoughts aloud'.[4]

Bereavement hallucinations are commoner still. A large 1971 study reported that nearly half of widows and widowers had post-bereavement hallucinations. Nearly 40 per cent had an illusion of feeling the presence of the dead spouse; about 13 per cent had auditory hallucinations; 14 per cent had visual hallucinations. More than 10 per cent had spoken to their dead spouse, and nearly 3 per cent had felt that they were touched by them.[5] I have discussed these phenomena in detail elsewhere,[6] but in the present context the brief comment 'So what?' will do. Some of these hallucinations do indeed seem to be similar in nature to some of the voices and other apparitions recorded in major religious works. But none of the psychologists' questionnaires says anything at all about the reality of the supposed apparition or voice. All that the studies do show (and they all show it) is that we seem to be rather better tuned in to frequencies other than the everyday than we tend usually to think.

Precisely the same goes for dreams. The Bible is full of dreams: those of Joseph, Daniel, Gideon, Nebuchadnezzar, Pilate's wife, and so on. If the Bible is to be believed, they seem often to have given very accurate information. The Bible gives God the credit for the accuracy, and nothing that science can say rubbishes that attribution.

Are these biblical dreams religious experiences? That's an unimportant matter of definition, isn't it? It is perfectly true that life-changing encounters with God himself don't appear to be mediated through dreams, but the course of many dreamers' lives has been changed by their emphatic conviction that God has dealt directly with them.

Science has exorcised very few of the occult mysteries of sleep, but one coven of demons has gone screaming out of the bedroom. These are the ones who sit on our chests, suffocating and squeezing, or, typically in the case of medieval nuns, sexually violating, in the gloomy twilight between sleep and waking. They have had various names in different cultures. To the Europe of the Middle Ages they were succubi and incubi (and some clerics spent their careers studying their taxonomy); in Newfoundland they are the Old Hag; in the Amazon they are the Boto; in some parts of southern Africa they are the Tikoloshe; and to contemporary medicine they are sleep paralysis.

When we are in the deep, dreaming, REM phase of sleep, our skeletal muscles are mercifully paralysed – presumably so that we do not act out our dreams, kicking and punching away the denizens of our rearing subconscious. Normally the paralysis ends when our unconsciousness ends, but sometimes they do not dovetail neatly. We half-wake to find ourselves unable to move, and give the paralysis a diabolical name.

So much for sleep: it seems to be used as a postbox by God, rather than a main stage. For maximum epiphanic impact he seems to prefer, at least in the Judaeo-Christian tradition, waking. For those apprehensions of divinity that use altered states of consciousness, over- or under-fizzing brains seem to do the job best.

We have noted already that many ways of rocking the natural equilibrium of the mind have been used to trigger those states of mind that are seen as particularly spiritually fertile. The brain's input can be diminished by more or less extreme forms of sensory deprivation – anything from mummification and flotation tanks to sitting on the top of a pillar in the Syrian desert for decades or having white noise piped into headphones. If the brain isn't fed enough, it chews on air and hates it. It will do anything to get a more entertaining diet. The energy of its convulsions can puncture the layers between worlds. To turn the idle brain's whining to hymn is the work of very great saints.

More commonly, though, the input is increased or its pattern changed. Extreme forms of masochism play a spectacular part in religious ceremony. In the Philippines, dozens of men celebrate Easter by allowing themselves to be scourged and crucified. In 2008 Filipino health officials solemnly advised that 'crucifixion is bad for your health'. They are right. On the day of Ashura, some Shi'ites remember the martyrdom of Ali by cutting themselves and beating themselves with chains. Ascetic Hindu holy men can be seen wandering the streets of India with skewers through their cheeks.

There are many things going on here, of course. These are not pure searches for the altered state of consciousness that pain can bestow. Guilt, a desire to show solidarity with a suffering religious figure, a lust for enhanced status in the community, the infectious intoxication of mass hysteria, and sheer bravado are no doubt mixed up in it all. But as the nails are driven in and the lashes fall, endorphins will be secreted, the thalamus will be flooded with impulses and the pain credited as righteousness in the mind of the victim. Later, exhaustion and dehydration will tip the physiology further in the direction of an altered state of consciousness, and the believer may start to float off.

There are purer, more effective ways of inflicting psychoactive pain. The monks are experts, and one of the first tools that a trainee ascetic learns how to use is hunger. Ascetics from all religious traditions have long taught that their prayers are more effective when they fast. By this, they at least partly mean that they *feel* closer to their god when their stomachs are empty. Partly this is psychological: most of them think, at some level, that the flesh is bad, and that god will smile the more broadly the more their turbulent flesh is subdued. But partly, no doubt, it is physiological. There will be pain, and hence afferent overload and endogenous opiates, and there may be low blood sugar and hence dizziness, which may feel like a physical drift to another place.

Popular, too, are sleep deprivation (reducing the ability of the brain to process the information it has received, leading to confusion-producing psychological disorientation and hallucination); dream induction (perhaps a form of self-hypnosis); adopting stressful postures (particularly in India); repeatedly dousing the head with water (again an Indian practice, said to be able to induce epileptiform activity);[7,8] and subjecting the body to extreme cold. The twelfth-century English hermit Godric of Finchale used to pray for hours in the River Wear, immersed to the neck, and practitioners of the Tibetan yogic disciplines of *tummo* and *kundalini* meditate naked or scantily clothed in arctic conditions. To read some of the Western books on *tummo*, you'd think that you couldn't cross a Himalayan glacier without picking your way gingerly through the crowds of cross-legged yogis navigating pulselessly through Nirvana. But it is a rare discipline. Some practitioners have mastered their physiology so completely that they can raise the temperature of their fingers and toes by an astonishing 8.3 degrees centigrade, and decrease their metabolic rate by 64 per cent (compared to 10–15 per cent in sleep and 17 per cent in simple meditation).[9] In 2008 the Dutch *tummo* adept Wim Hof took the world record for surviving more or less naked in a tub of ice: one hour and thirteen minutes.

A practitioner of the yogic discipline of tummo
meditates in the snow, almost naked.

Kinder ascetics may change the tempo at which the mind walks. Mere repetition can lull the brain into a trance state in which anything spiritual is graspable and anything is bearable (soldiers marching in step to a fife and drum will go for hours and miles longer than troops allowed to shamble along at their own pace). The lilt of Gregorian chant, the endless turning over of a Tibetan mantra, the saying of the rosary, the stultifying tedium of house music in a nightclub, the umpteenth fireside rendering of a well-known bardic poem, the New Age relaxation CD of waves crashing on a beach, a children's lullaby, and a Buddhist meditation session in which you count your breathing up to ten and back again for an hour, all have identical effects on the brain.[10] The brain gets quickly confident that it is not going to have to deal with any novelty. It knows that it can go off and fly.

Rhythmic movements of the body have a similar effect, and for similar reasons (although if the activity is strenuous the physiological picture might well be joyously complicated by happy-making compounds such as serotonin). The dancer at that house music club is doing the same thing as the marathon runner, the rosary counter and the charismatic Christian stamping her feet to the song 'You took the shackles off my feet so I could dance'. If you go to the Western Wall in Jerusalem you will see black-coated Orthodox Jews praying. Many will be nodding vigorously and swaying. Their movements mean many things. They are saying that their bodies worship, as well as their minds, and that the body is happier to let the mind concentrate if the body itself has something to do. But on another level those pious Hasidim stand squarely in the tradition of the ecstatic clubber and the dancing shaman.

We came as a species out of Africa, and if we want to see ourselves clearly we need to look at our reflection in the waterholes of southern and eastern Africa. Wherever there is the

Turkish dervishes whirl round and round. Dancing, and particularly dances with repetitive or disorientating movements, have long been used by adepts in many spiritual disciplines to usher them into altered states of consciousness.

thinnest of tunnels through that porous membrane separating our world from others, a bushman of southern Africa will have found it aeons ago and wriggled through, coming back to tell the tale and to show his apprentice the route.

They dance, starved, through the day and the night to the throb of drums and the murmur of old songs. The firelight flickers and breeds ghosts. When exhaustion and pain come, they are happy because they know that the spirits are close. Dehydration makes the vessels in their nose fragile. The vessels burst, spraying the sand with blood, and they dance on, round and round, supersexually aroused because of the imminence of a non-sexual consummation. The pains are like transfixion with a char-tipped spear. At the edge of their vision the dancers begin to see the dark river through which they have to swim. Then suddenly they are teetering at the top of the bank and plunging in, and swimming through a speckled tunnel to the far bank, where they are met by the dead.

Sometimes, although not particularly in Africa, other worlds have permanent or semi-permanent ambassadors in this world. Their embassies are usually human bodies. These are the oracles, from which the ancient world often, and sometimes disastrously, took its cue. The Dalai Lama's exiled government of Tibet includes the Nechung Oracle, a spirit guide occupying the body of a man. The Dalai Lama's relationship with the oracle is very businesslike and matter-of-fact. The oracle is consulted, says the Dalai Lama, because he has always found the oracle to be correct:

This is not to say that I rely solely on the oracle's advice. I do not. I seek his opinion in the same way as I seek the opinion of my Cabinet and just as I seek the opinion of my own conscience. I consider the gods to be my 'upper house' . . . Nechung has always shown respect for me . . . At the same time, his replies to questions about government policy can be crushing. Sometimes he just responds with a burst of sarcastic laughter. I well remember a particular incident that occurred when I was about fourteen. Nechung was asked a question about China. Rather than answer it directly, the Kuten[11] [the medium in whose body the spirit lived] turned towards the east and began bending forward violently. It was frightening to watch, knowing that this movement combined with the weight of the massive helmet he wore on his head would be enough to snap his neck. He did it at least fifteen times, leaving no one in any doubt about where the danger lay.[12]

The Nechung Oracle deserves a mention here precisely because the phenomenon, whatever its real nature, is spectacular. But spectacular examples risk clouding a quieter, yet in its way even stranger phenomenon: the fact of occasional transport to other places while sitting at one's desk or on a bus.

C. S. Lewis wrote of a childhood experience – in fact a memory of a memory – that coloured his whole life:

> As I stood beside a flowering currant bush on a summer day, there suddenly arose in me without warning, and as if from a depth not of years but of centuries, the memory of that earlier morning at the Old House when my brother had brought his toy garden into the nursery. It is difficult to find words strong enough for the sensation which came over me. Milton's 'enormous bliss' of Eden (giving the full, ancient meaning to 'enormous') comes somewhere near it. It was a sensation, of course, of desire; but desire for what? Not, certainly, for a biscuit-tin filled with moss, nor even (though that came into it) for my own past . . . Then the curtain fell, but the world was ever after different . . . [I]n a certain sense everything else that had ever happened to me was insignificant in comparison.[13]

Lewis was not alone. He was not even unusual, although he was, of course, unusually articulate. But one's capacity to experience is not a function of one's ability to express the feeling.

Strip out all those febrile, dramatic drivers to altered states of consciousness, so beloved of the neurophysiologists and the ethnologists, and you still have the almost universal conviction that we're just 'passing through'; that there is a real world on the other side of an invisible veil; that this is not all there is; and that there are moments of breakthrough – moments unassisted by drugs, drums, congregational enthusiasm, tiredness or even conviction. It is as if our brains are antennae which sometimes pick up broadcasts from another place – a place which is strangely familiar, a place that seems more like home than home itself.

CHAPTER 10

Turning On and Tuning In: Brains as Antennae

Turn on, tune in, drop out.
> (Timothy Leary, at the Human Be-In,
> Golden Gate Park, San Francisco, 1967)

Hallucinogens and other means of inducing altered states of consciousness work by temporarily 're-tuning' the brain to pick up frequencies, dimensions and entities that are completely real in their own way but that are normally inaccessible to us.
> (Graham Hancock, *Supernatural: Meetings with the Ancient
> Teachers of Mankind*, London: Arrow, 2006, p. 346)

There is nothing particularly strange about believing in gods. In fact belief in gods in human groups may be an inevitable consequence of the sorts of minds we are born with in the sort of world we are born into.
> (Justin Barrett, *Why Would Anyone Believe in God?*,
> Lanham, MD: AltaMira, 2004, p. 91)

Remember Huxley's idea of the brain as a 'reducing valve' – the notion that the brain selectively reduces to a manageable dribble the vast stream of data flooding through the sensors?[1] The brain scoops up the dribble and uses it to create a highly – perhaps catastrophically – edited version of reality. It is neurological nonsense to say that we perceive 'reality'. We don't even know how good an approximation to reality our own model is. All that we can say is that it is our own – rather than being an objective reflection of what is out there – and that it is bound to be sadly lacking.

Some drugs and some physiology-altering practices seem to be able to slacken off the reducing valve. The result is that we let more information gush into our heads. The model of the world that is then built is different. Whether or not it is more accurate than our usual, workaday model is difficult to say. But it does seem to be bigger. The popular expression 'mind-expanding drugs' might not be that far from the mark. Reports of NDEs often use language that suggests that something similar is happening there.

'My presence fills the room,' wrote Josiane Antonette, fairly typically, of her NDE. 'And now I feel my presence in every room in the hospital. Even the tiniest space in the hospital is filled with this presence that is me. I sense myself beyond the hospital, above the city, even encompassing Earth. I am melting into the universe. I am everywhere at once.'[2]

Yet this feeling of ubiquity is a special example of a more general observation – the feeling of being intimately connected to everything else. This is expressed sometimes in the language of Absolute Unitary Being,[3] but probably more often in the language of broadcasting. It is as if being mystical means that you can receive on a much wider range of bandwidths than usual.[4]

Here is the testimony of a user of the dissociative drug ketamine (K):

Is this your passport home? [In my humble opinion] K is a tool for rapidly and safely accessing hyperdimensional realities (usually [zero] gravity) which exist vibrational frequencies away from baseline beta frequencies.[5] K allows you to control the frequency your body-mind ([Central Nervous System]-DNA) is tuned to and allows you to operate at much higher vibrational frequencies. The feeling when coming back Earth-side is that every molecule has been recharged.

Think of a TV set tuned only to the sports channel. Only picking up sports-related info. Now, give that person a device to tune the TV into other frequencies. Wow, other info. Food, weather, sit-coms, news, religion, porn, nature, music . . . get the picture? We all have this reality-tuning ability within us. Yessirree, K mimics the effects of an endogenous neurotransmitter, lovingly called angeldustine. Secreted at moments of extreme stress and/or spiritual ecstasy. This chemical explains the OBEs, astral-projections, NDEs, other-worldly experiences, extra-terrestrial contacts etc.[6]

Rick Strassman, the psychologist who used DMT on experimental subjects, spoke similarly about the effect of DMT:

No longer is the show we are watching everyday reality, Channel Normal. DMT provides regular, repeated and reliable access to 'other' channels. The other planes of existence are always there. In fact, they are right here, transmitting all the time. But we cannot perceive them because we are not designed to do so; our hard-wiring keeps us tuned in to Channel Normal. It takes only a second or two – the few heartbeats that the 'spirit-molecule' [DMT] requires to make its way to the brain – to change the channel, to open our mind to these other planes of existence.[7]

The K-user suggests that the version of reality that one gets with this K-retuning is hugely more satisfactory than the one we usually have to be satisfied with, and paints a picture of what we might be able to do if (to change the metaphor) the valve were normally rather slacker than it is.

With practice, using K you can begin to give your body-mind very specific instructions, which it will perform with robotic precision. Not only stand up, balance, take three steps forward, press the Hi Fi power switch in, and so on. But also very intricate movements that demand incredible fine neuromotor control. I mean complete a jumbled Rubik's Cube in under a minute, play the violin like Paganini, shoot that arrow better than Robin Hood, type up an award-winning magazine article. Here is a new market. Jack a tab and play like a pro. . .[8]

Well, I wonder. All the research suggests that although he might have been able to play like Paganini in the world to which K took him, and was no doubt applauded thunderously there, it wouldn't have sounded like Paganini in London. In London you would have heard incoherent scraping. But perhaps that would be the fault of your wrongly tuned ears. William James, one of the most articulate of men, thought, when he was on nitrous oxide, that he was having insights of immortal genius. And, again, perhaps he was a genius by the standards of wherever he was at the time. Some of his jottings survive:

What's mistake but a kind of take?
What's nausea but a kind of -usea?
Sober, drunk, -unk, astonishment.
Everything can become the subject of criticism –
How criticise without something *to* criticise?
Agreement – disagreement!!
Emotion – motion!!!!
By God, how that hurts! By God, how it *doesn't* hurt!
Reconciliation of two extremes.
By George, nothing but *oth*ing!
That sounds like nonsense, but it is pure *on*sense!
Thought deeper than speech. . .!
Medical school; divinity school, *school*! SCHOOL!
Oh my God, oh God; oh God!

James said himself, 'The most coherent and articulate sentence which came was this: There are no differences but differences of degree between different degrees of difference and no difference.'[9]

Perhaps this really is how poetry should be written.

If it pleases you, you can move straight on from the K-user, Strassman and James to assert that there are irresistible parallels with Jesus' talk about the kingdom of heaven being close at hand, and about the kingdom being a topsy-turvy place in which children are the wise ones and in which, perhaps, artless gibberish counts as poetry and apparently random caterwauling knocks Paganini cold. If it pleases you, you can use these parallels as the basis for an assertion that the 'kingdom of heaven' is code either for a simple altered state of consciousness, or for the world that is genuinely accessed while one is in an altered state of consciousness. But to my mind it simply won't do. To maintain the thesis you have to put a line through most of the other things that Jesus did and said. In particular you have to dispense with the whole notion of the resurrection, which does rather seem to be the main point of the whole story.

But a faithful Christian can surely acknowledge that a widening of the range of receivable frequencies might be interesting, and at least potentially useful and helpful. If the K-user doesn't actually sneak in the course of his trip past the watchful eye of St Peter, it doesn't at all mean to say that he hasn't been close to the gates. Still less does it mean that he's wandered in forbidden, demonic realms. The kingdom is likely to be a big place. On first principles one might be rather more surprised if it existed only on one narrow bandwidth than on many. The parallels between the mystical experiences traditionally endorsed by the Church as genuine and those reported by the users of other, less doctrinally kosher routes are just too close to be dismissed.

While the mystical experiences accepted in the Judaeo-Christian tradition are certainly not identical to those from other traditions and none, and there are some important differences of emphasis, there is also an important degree of convergence. We can quibble about the differences, but the common point is so obvious that it is easily missed. While on Strassman's 'Channel Normal', almost everyone who has ever lived has thought that there is some sort of God or gods. Atheism is a position held by so few people that, if we're just counting heads over the aeons, we can forget about it. To a first approximation, there are not and have never been any human atheists. If you don't like this generalisation, you'll like the next one even less: no one, but no one, is an atheist when he's off his head on LSD or magic mushrooms, or in the throes of a shamanic trance. If Richard Dawkins twiddled his dial and received a slightly broader range of bandwidths, he'd never have written *The God Delusion* and would now be living in a much smaller house.

It might sensibly be objected that I have moved on from talking about mystical experience (the subject of this book, and presumably the reason for the reader buying it) to talking about religious belief (a wholly different thing, which the reader has not bargained for). And I entirely agree that they are not the same thing, and that it is important not to conflate them. One can have one without the other. But nonetheless there is a connection. A jaw-dropping spiritual epiphany is likely to import a set of beliefs, or at least brush out others. The new beliefs might very well not coincide with the catechism of any conventional faith, but that is quite another matter. As for beliefs without experience: well, some strait-laced Protestants might say that experience of any kind is suspect, and viciously excise experience from their own devotional lives, but by and large religious believers are also religious experiencers.

Because of the close association between experience and belief, work on belief might be able to tell us something about why our antennae are tuned the way they are; why they are apparently so capable of picking up, with minor adjustments or none, bandwidths other than Channel Normal; whether mystical or religious people are odd, or whether the really odd people are the non-religious.

It is wholly misleading to talk about 'tuning', many say. We don't need to invoke any transmitters 'out there'. There are perfectly good naturalistic explanations for the beliefs that we hold.

There are two types of naturalistic contention. The first is that religion and mysticism themselves are generated by Darwinian natural selection. That possibility is discussed in Chapter 11, and I have also dealt with it in detail elsewhere.[10] The second is that the particular beliefs we hold are inevitable by-products of our biology. The two contentions of course overlap,[11] but it is important to recognise that they are analytically distinct. It is with that second question that we are mainly concerned here.

There's no issue of 'tuning', say Newberg and Waldman. 'The human brain is really a *believing* machine.'[12] '[W]e are biologically inclined to ponder the deepest nature of our being and the deepest secrets of the universe.'[13] The most basic clause of the brain's constitution is the cognitive imperative.[14] There's an 'almost irresistible, biologically driven need to make sense of things through the cognitive analysis of reality'.[15] 'It's our duty to understand,' wrote Rupert Brooke, 'for if we don't, no one else will.'[16] That's the duty that the brain feels.

Wielding this imperative, Newberg and D'Aquili get to work explaining belief. They see myths (amongst which they count religious beliefs) as a product of this imperative. Myths, they contend, are manufactured (originally by people with vastly hypertrophied parietal lobes)[17] to make sense of the otherwise incomprehensible: death, suffering, creation and so on. They suggest that all myths can be reduced to a simple framework: (a) an existential concern – such as why there is evil in the world; (b) a framing of that concern in terms of apparently irreconcilable opposites (such as heaven/hell, life/death, etc.), and (c) a mythological reconciliation of the opposites. To look at myth as a set of binary opposites is straightforward Levi-Strauss.[18] Levi-Strauss has fallen out of vogue, and great was the fall. His was a nice idea, but too neat to be true. The anthropological field reports just don't support him. It seems to me that Newberg's and D'Aquili's explanatory framework has to fall with him.

Laurence McKinney proposed – in a twist on Freud – that we're religious because we've forgotten lots of important things that we knew in our childhood. Our adult brains desperately want answers to questions like 'Where did I come from?', can't retrieve the answer (it's presumably jumbled up amongst memory files labelled things like 'My first pair of trousers' and 'The death of the guinea pig') and, unable to cope without an answer, invent God as the Answer To End All Answers.[19] Then there is Freud himself, whom we have met already.[20] For him, adult life is a crashing disappointment. We all look back with aching nostalgia to the Golden Age of blissful completeness and oneness with our parents and with the world in which, our deceitful memories tell us, we lived as infants. We can't live with the idea of complete estrangement from it, and accordingly generate God both to describe the Golden Age and to take us back there. Freud and McKinney can enlist some illustrious literary worthies on their side. The title of Wordsworth's great poem 'Intimations of Immortality from Recollections of Early Childhood' is as good a précis of the Freudian thesis of religious infantilism as you could get.[21,22]

There are some terrible difficulties with these ideas. Perhaps the most obvious is that the very youngest children we can study seem to be emphatically religious. You get *less* religious as you get older, and then often more religious later in life, in what seems, from a religious perspective, to be a genuine second childhood.

Religion, it has been vigorously contended, is a systematised anthropomorphism.[23] We create God in our own image, or in the image of an ideal adult. Thus Piaget would say that children start off by thinking that adults are omniscient and omnipotent, but gradually realise that they are not: that their idols have feet of clay. This is unbearable, and belief in an omnipotent and omniscient God is the result. God does what those fallen idols should have done. It's Freud and McKinney warmed up and re-served with a bit of modern psychological salad.

But it's wrong. One of the central pillars of Piaget's thesis was the assumption that young children think that the natural world is created by people: that some human drew the blueprint for badgers, bolted snakes together, and can make rain come when it's dry. But they don't. When British schoolchildren were asked where rocks, plants and animals came from, and were given the three options of 'people', 'God', or 'no one knows', they were seven times more likely to say 'God' than 'people'.[24] Children are intuitive creationists and intuitive theists.[25]

They also, quite naturally, without any brainwashing in the Sunday school or the madrasah, credit God with certain other characteristics in addition to creatorship.

When my son Tom was three years old, I showed him a matchbox. He knows that a matchbox normally has matches in it, but I had taken the matches out and put a small model duck inside. I asked him what was in the box. 'Matches,' he said. I then showed him the duck, and closed the box again. I then asked him to imagine that his mother came into the room. 'What would she think was in the box?' I asked. 'A duck,' he said, without hesitation.

Assuming that his cognitive development is normal, when I try the experiment again when he is five, and ask him what his mother would think is in the box, he will then answer, 'Matches.' He will have realised that people can hold beliefs that are untrue. But here's the point: children of *all* ages, when asked what God would think was in the box, would answer, 'A duck.' They believe that God can't be fooled, although they know that their mother can.[26]

Rebekah Richert got children, aged three to eight, to look through a slit in a darkened box. 'What do you see?' she asked. 'Nothing,' came the reply. She then shone a light into the box, revealing a wooden block. The light was then turned off. The children were told that cats had special eyes and could see in the dark. She then produced monkey, human and cat puppets, and asked what each, and God, could see. Only God and the cat could see the block, said the children.[27]

God, in short, isn't a strictly anthropomorphic projection.[28]

Some have suggested, though, that he might be a by-product of the almost manic human tendency to see *agency* everywhere.[29] Just look at how we treat computers, for instance. If our computer misbehaves, we attribute to it a malicious state of mind. We swear at it and hit it. If we trip over a chair, we kick the chair, assuming, at some level, that it is out to get us. If the sun shines and our mood is bright, 'the universe' is on our side.

Justin Barrett writes:

Part of the reason people believe in gods, ghosts and goblins . . . comes from the way in which our minds, particularly our agency detection device (ADD) functions. Our ADD suffers from some hyperactivity, making it prone to find agents around us, including supernatural ones, given fairly modest evidence of their presence. This tendency encourages the generation and spread of god concepts and other religious concepts.[30]

It is easy to see the selective advantage both of having an ADD, and of having it tuned in the direction of over-sensitivity (what Barrett labels 'HADD' – hypersensitive agent detection device). If our ancestors were wired to think that the rustling in the grass might be a malignant agent – say a sabre-toothed tiger – they were more likely to have survived to be our ancestors than if they had simply shrugged and said, 'Nah, it's just the wind.' A hundred paranoid mislabellings of innocent rustlings as deadly, spitting lumps of fur and sinew are far less costly than a single mislabelling of a hungry cat as a rustle.

There is presumably some correlation between HADD and theory of mind (TOM) – the ability to think your way inside another person's head – although the exact nature of the relationship is obscure, at least to me. HADD says, 'That's an agent,' and TOM says, 'Yes, but if it isn't showing itself, it must be more afraid of you than you are of it, and so if you keep on looking at it, it will think that it's simply not worth trying to have you for dinner.'

The cliché is perfectly true: churches tend to be full of old women. All the studies have shown it. Across all religious traditions, in fact, women are more committed and involved than men. This, speculates Barrett, might be because of that supposed relationship between HADD and TOM.[31] By and large, women have more TOM than men. It has been truly said that all men are to an extent autistic and autism, of course, is characterised by an inability to read the minds of others: it is a disorder of TOM. As one gets older, one hopefully, and probably actually, gains sympathy, experience and general social intelligence – qualities that can boost one's existing TOM, or make up to some degree for truncated TOM. Perhaps that's why people get more religious when they get older.[32]

Where does all this leave us, and how, in particular, does the

HADD hypothesis for supernatural entities relate to the three-year-old with the matchbox?

It is important to realise what HADD can and cannot do. HADD explains plausibly why, when we hear a bump in the night, we shout, 'Ghost!' But go far from bumps in the night, and the explanatory power of HADD quickly founders. You simply don't need God, or anything like him, for almost anything with which the Upper Palaeolithic mind would have been concerned. Not only do you not need him, but, as we will see in Chapter 11, and as the reductionists like Dawkins agree, religious belief and ritual exact a crippling price for no very obvious benefit.

But even if HADD can conjure a god, it cannot give him the qualities that those children so emphatically attributed to God. If HADD makes God psychologically inevitable, it will not make him omniscient or omnipotent. Still less will it give him many of the other characteristics that the great religions say he has. Most notably, it won't make him moral.

Barrett concludes:

> Developmental evidence suggests that children have built-in biases that encourage them to understand and believe (at least in some rudimentary sense) in super-knowing, super-perceiving, immortal, super-powerful creator gods. God concepts (such as those in Christianity, Islam, Judaism and some forms of Hinduism) that have these properties enjoy some transmission advantages over other god-concepts. Thus *once introduced into a population*, God concepts hold strong promise to spread rapidly and gain tenacious adherents. The histories of Christianity and Islam illustrate this claim.[33]

Those italics are my own, but they are crucial. To give biologically coherent reasons for dissemination is a million miles from accounting for origins. Natural selection would have sprayed metaphorical napalm on the seeds of religion.[34]

HADD, too, can't offer even a tentative explanation for the existence of those 'biases'.

So, going back to those antennae, we simply don't have a credible naturalistic explanation for our possession of antennae that are tunable to a God channel; still less do we have such an explanation for why they are tuned so neurotically to that wavelength. Antennae that will pick up the signal of breeze stirring the bamboo, and translate it to 'tiger', simply aren't up to the job. You need another model entirely for the God channel. And so, of course, we have no explanation for the apparent fact that one can fumble pharmacologically, meditatively, drunkenly, prayerfully, masochistically and in all sorts of other ways with the dial to pick up other broadcasts from the same general direction, but to which, in our waking modes, we are not normally tuned.

Of course the present absence of a naturalistic explanation doesn't at all mean that there is no naturalistic explanation. But it does mean that the supernaturalist needn't be embarrassed to share a podium with Richard Dawkins.

CHAPTER 11

Religious Experience and
the Origin of Religion

Yes, it may all be due to a few misfiring neurons, perhaps an
extra dollop of neuropeptide or whatever, but the fact remains
that humans have an overwhelming sense of purpose. As a species
we are strangely comfortable to find ourselves embedded in a
teleological matrix.

(Simon Conway Morris, *Life's Solution:
Inevitable Humans in a Lonely Universe*, Cambridge:
Cambridge University Press, 2003, p. 313)

While celebrating mass in Naples on 6 December 1273, Thomas
Aquinas, the great architect of scholastic thought, had some
sort of ecstatic epiphany. He was never the same again. His
secretary urged him to get back to work. There were some
daunting writing projects afoot. 'I cannot,' said Thomas. 'All
that I have written seems like straw to me.' He died in 1274,
while expounding the Song of Songs.

To talk to some Christian ultra-conservatives, you'd think
that ecstatic experience and true religion were mutually exclusive. For them, to *feel* anything is to have listened illegitimately
to the promptings of the world and/or the flesh and/or the devil.
True religion, they say, is obedience to the Word of God as
handed down, and that Word is immutable. It doesn't change
with your feelings about it, or with the weather, or with the
state of your digestion. Therefore feelings add a self-created
gloss to your view of what the Bible says, and should be stamped
on. This position is riddled with non-sequiturs and ironies –

not the least of which is that reason, of all things, is appealed to as the opponent of feelings. The irony is not so much that reason and feelings are set up as opponents, but that the ultras think that they, who so loudly eschew so much of what scholarship tells us about the world, are the only reasonable ones.

Some forms of knowledge are more intimately connected to experience than others. I spent months chewing over a Zen *koan* – a paradox incapable of solution by rational thought. My left brain was humiliated. It refused to admit defeat. But eventually it got tired, and that gave another sort of knowledge a chance. My mind was suddenly and wonderfully shoved off the rails that normally prescribed its route. For the first time since I was a child, it could go anywhere. And there were new experiences, and very old experiences revisited. There was the

'The Sorcerer' – an Upper Palaeolithic horned human–animal hybrid from the wall of the Gabillou cave, Dordogne, France. One theory is that such figures represent shamans on the point of transformation into 'spirit animals'. Another sees them as representations of the therianthropes encountered in the other worlds accessed by entheogens or other forms of shamanic practice.

simple buzz of freedom, as well as the sights, smells and sounds of newly accessible places.

Whatever the exact relationship between religion and experience, most people, whether in the pews or the laboratory, agree that they are crucially and fundamentally connected. But the nature of that connection is highly controversial. Religion is plainly not merely the experiences that we call religious. Belief and ritual are also components. How those components relate is obscure. Here is one suggestion.

An Upper Palaeolithic hunter stalked a deer for many days. He was hungry, dehydrated and exhausted. We know that those conditions can generate altered states of consciousness. He entered such a state and saw what the San bushmen of southern Africa saw (until they were hunted to extinction by white settlers) when they danced around their fires until blood poured out of

The famous giant deer at Lascaux, Dorgogne, France. Only its fore-quarters are visible, suggesting that the rest of the animal is in the world on the other side of the cave wall. The deliberate dots may represent entoptic phenomena encountered during the shamanic journey to that other world.

their nostrils. He saw geometric patterns – perhaps in a lattice. He touched the thin skin that separated him from another world. It trembled at his touch. He passed through a tunnel into that other world. There he saw therianthropes – animal–animal and animal–human chimaeras; he saw spirit animals and the face of his dead father.

He came slowly back to the forest. The deer, puzzled, was watching him from a few paces away. He raised his flint-tipped spear and flung it into the deer's chest.

After he had eaten some of the deer's liver to revive himself, he set to thinking. He had no choice but to think. Human beings can't help it. They have to make sense of things. They are hopeless at living with uncertainty. When a new experience comes along, the left hemisphere will knock up a pigeonhole to put it in. The urge to file seems to be a lot greater than the urge to file accurately. As long as an experience can be forced into some explanatory category or other, the left hemisphere is happy. As we have seen, once an experience is fitted into a particular analytical framework, it is very hard to persuade the left hemisphere to reshape the framework so that it doesn't rub so uncomfortably. As we have also seen, humans have an inbuilt tendency to attribute agency to things, and agents are the most intuitively satisfactory candidates for causes. The brain loves causes; it loathes causelessness, and can't live with it. The cognitive imperative insists that the brain makes a finding of causation. It is not enough to say, 'I don't know what caused that. The evidence isn't sufficient.' The evidence, if necessary, is artificially enhanced until the brain has convinced itself that a finding of causation can confidently be made.

The hunter had to make sense of what had happened to him. He hadn't entered the 'spirit world' before, nor had a deer stood and offered itself to him. The connection between the two events was inescapable. The 'spirits' had caused the deer to stand there. They were on his side. Perhaps his father was the head spirit. Perhaps the deer-headed man he had seen was a spirit in the process of shape-shifting, about to leap through into the forest and manifest itself as the deer he had killed. Perhaps, if he could get back into the spirit world, he could persuade other spirits to change into deer, which would solve the problem of his tribe's food for the winter.

By the time he had got back to his tribe, these thoughts had crystallised into belief. Theology was born. It was plain, too, that it was important to keep on the right side of these potent spirits. They were powerful. What they gave with one hand they could doubtless take away with the other. It would be wise to propitiate them with food and respect. So ritual began. The neuronal firing involved in repetitive ritual entrenched the beliefs.[1] (Rituals themselves, regardless of their relationship to belief, become addictive. We all know cradle Catholics who, though they contemptuously reject all the beliefs of the Church, are to be found genuflecting obediently every Sunday.) Once neuronally entrenched, society took over part of the policing of the new orthodoxy. Disbelief meant ostracism. At this point, natural selection began to take an interest in the preservation of the belief: ostracism would disastrously diminish the chances of gene transmission to the next generation.

But nonetheless, a belief system based on an old memory by one hunter of one glimpse of another world was vulnerable. That sceptical, holistic right hemisphere might, in the absence of corroborating experience, generate disloyalties. And natural human curiosity was in play too. 'We would like to see the spirits too,' said the other tribe members. 'I would like to see my dead husband again,' said a widow. And so, deliberately or accidentally, the conditions that generated the original altered state of consciousness were rediscovered. You didn't have to hunt for five days without food: you could dance round the fire for twelve hours. You didn't even have to dance, in fact: you could eat those red and white mushrooms growing by the lake. You saw slightly different spirits if you did that, but the general effect was the same. And so, added to the rather dull entrenching rituals of propitiation and story-telling were the exhilarating entrenching rituals of a fungal sacra-ment and a liturgical dance. You can go straight from that Cro-Magnon campfire to High Mass in St Peter's.

It is an intoxicating thesis, and it has intoxicated. But it doesn't work. It begs the very questions it purports to answer.

Many of its elements are true. Our brains do indeed seem compelled towards self-transcendence. If for 'God' we read 'self-transcendence', Thomas Aquinas was neurologically spot on when he wrote, 'It seems that the existence of God is self-evident. Those things are said to be self-evident to us the knowl-edge of which is naturally implanted in us.'[2] The biologist E. O. Wilson agreed: 'The predisposition to religious belief is the most complex and powerful force in the human mind and in all probability is an ineradicable part of human nature.'[3] It is terribly, terribly difficult to be an atheist, which is why almost nobody, until very recently, has been.

Atheism is so unnatural that a young, faithful atheist must guard his faith very carefully. It will be assaulted on all sides by almost everything he sees in the world and in his own head. Justin Barrett gives atheists some wise advice.[4] First: let evolu-tion be your explanation for absolutely everything. In particular, let your natural tendency to see agents everywhere be satisfied with evolution as the agent of everything. (Does that sound familiar, readers of The God Delusion?) Second: stay out of foxholes; there are famously no atheists there. Avoid all really urgent situations: your God-invoking instincts are likely to spring out embarrassingly at the sight of a flashing blue light. Third: stay urban. Avoid the countryside. Allow yourself to see only human agents. The sight of a lightning strike, a bird pulling up a worm, or buds peeping insolently out might be the thin end of a dangerous theistic wedge. Fourth: avoid second-hand accounts that might be construed as evidence for God. It is probably best to avoid all religious people, but if that can't be managed, be so promiscuously pluralistic that you never stop to think about whether the experiences of others might indicate something true. If lots of experiences clamour loudly for attention, you won't be able to evaluate reflectively any one in particular. If you do, beware: there have been many sad cases in which mere academic evaluation has ended up at the altar. And finally, remember that over-analysis is both an effective prophylactic against any sort of belief, and a treatment for it. If you are unwise enough to go into a church, spray yourself liberally with the sort of maxi-strength cynical reductionism that you know would knock out belief in absolutely anything, including the second law of thermodynamics, and you'll emerge unscathed.

Barrett, in fact, thinks that religious experience of a mystical kind isn't convincingly linked to the prevalence of religious belief: 'Belief in God or gods does not arise because of peculiar brain states or psychological abnormalities.'[5] That belief, he says, is the norm. And I agree, but with two important caveats. First: there is nothing pathological about the brain states associated with mystical experience. Those states themselves are absolutely normal. A large proportion of us have them at various times, and an even larger proportion could. Mysticism is as much a part of being a human as being bipedal. Indeed, it is rather similar. If we try to get around on all fours, we make very slow progress. If we try not to be mystical, we are badly handicapped in some very significant ways. And second: while a mystical experience might not be the cause in a particular individual of belief in a God or gods, that is not to say that mystical experience was not the catalyst for religious belief itself in the species, or of the particular religious system in which the individual has come to believe. We are concerned at the moment about how religion first came to be injected into the aboriginal *Homo sapiens*.

There is no convincing evidence of religion in other species of human – all of which are now extinct. The Neanderthals apparently buried their dead, but this does not necessarily indicate that they thought there was another world to which the dead had gone. It might have been done to keep the smell of decomposition out of the cave, or to avoid the unwanted attention of sharp-toothed scavengers. Controversy rages over the famous Neanderthal 'flower-burial' in Shanidar, Iraq. A man lies in a grave dusty with the pollen of spring flowers. Were the flowers early grave-goods, intended to beautify the deceased's room in the land of the dead? Were they thrown there sentimentally by a bereaved widow or child? In either case the find would be hugely significant. It would indicate a belief that the man had a significance other than the merely physical: that something about him – even if it was only a memory – lived on once his body was carrion. In either case you have the building blocks of a theology. But the truth might well be more prosaic: the burrowing of the local rodents can distribute pollen in the way seen in the grave.

What everyone agrees about is that religion and symbolic thought explode into the world at the same time. The relationship between them is bitterly contested, as is the speed of the explosion.

Was religion an inevitable by-product of the neural hardware and software that produced the ability to express the world in symbols? The ability to think in symbols allows you to generate in your head and to test any number of different models. It allows you to ask, 'What if?' and that is the question that sets *Homo sapiens* apart (so far as we know) from all the other animals that have existed. It is obviously immensely advantageous to be able to ask it. Instead of having to learn your lessons in the brutal school of savannah experience, populated as it is by hostile claws and spears, you can sit in camp having mental dry runs. They are much faster, safer and more energy efficient than real trial and real error. Perhaps, it is speculated, the notions of gods, other worlds and self-transcendence were simply amongst the many models of reality incontinently spawned by the newly symbolic brain.

The dramatic flowering of human symbolic representation:
The Venus of Brassempouy, a mammoth-ivory figurine from
south-western France. At 25,000 years old, it is one of the earliest
known detailed representations of the human face.
(Musée d'Archeologie Nationale, France)

Language is one of the obvious means of representing the world symbolically. It is also probably one of the earliest. Lots of people think that language was the original symbolic faculty, rather than being the fruit of a more basic neurological facility to symbolise. If that's so, and religion is a by-product of the waves of 'What-iffing' that characterised early human symbolising, then it might be literally true that humans spoke God into existence, rather than the other way round. We return shortly to the question of how likely this is.

What about the speed of the symbolic explosion?

The conventional position has broadly been that anatomically modern *Homo sapiens* – skeletally indistinguishable from you and me – first appeared in Africa about 100,000 years ago. All the neural hardware was in place to allow this early man to do everything that we do. In terms of processing power, memory and general cognitive apparatus, there was nothing to stop him occupying the Oval Office with distinction. But, very strangely, he doesn't symbolise. He doesn't bury his dead. He's not religious. He behaves just like an ape, when there's no obvious need to. Surely natural selection would have been protesting loudly at such a waste of splendid neural raw material. But, oblivious to those Darwinian proddings, he squats in the East African dust for about 50,000 years, eating roots and being eaten by animals far less intelligent than himself.

And then, almost overnight, about 45,000 years ago, he changes. The change is first seen in Europe, but we don't know if the behaviourally archaic man migrated from Africa to sit in the European drizzle for millennia, squandering his biological potential until his epiphany came in Europe, or whether the newly enlightened, behaviourally modern man that we first detect in Europe was made modern in Africa and then headed north. The change looks dramatic. It is the dawn of the Upper Palaeolithic, and the dawn of the real us. The record seethes with symbolism. Apart from Shanidar, the most dramatic earlier candidates for symbols are a few ambiguous scratches on pieces of bone. Now we see elaborate jewellery, combs, musical instruments. And religion – religion everywhere. Also pictorial representations of a very strange and sophisticated kind, to which we will return shortly. Ian Tattersall writes:

It is no denigration at all of the Neanderthals and of other now-extinct humans – whose attainments were entirely admirable in their own ways – to say that, with the arrival on earth of symbol-centred, behaviourally modern *Homo sapiens*, an entirely new order of being had materialised on the scene. And explaining just how this extraordinary new phenomenon came about is at the same time the most intriguing question, and the most baffling one, in all of biology.[6]

Indeed. It is a question to which I insolently attempt a tentative answer in *The Selfless Gene*.[7]

This conventional view – the idea of an overnight symbolic efflorescence – has recently taken some knocks, although it still has many powerful proponents.[8] Critics point to several isolated and apparently independent outlying incidences of earlier behavioural modernism in Africa. It is certainly true that the Mesolithic has not been documented as thoroughly in Africa as it has in Europe: there is a lot more to find in Africa if one looks. Perhaps symbolic thought flared up independently in several different places, but only really caught the neurological brushwood alight in Upper Palaeolithic Europe. Perhaps there were established communities of behaviourally modern humans in Africa well before the Upper Palaeolithic; perhaps they seeded the European Upper Palaeolithic; or perhaps the African early modern communities did not have the necessary critical mass, and died out, leaving symbolic thought to be reignited independently. Perhaps there were several Gardens of Eden throughout Africa, the Near East and Europe, only one or a few of which survived to populate the world with truly sapient *Homo*.

All that said, though, it seems unlikely that the overwhelming impression in the record – of fully wired, anatomically modern but behaviourally archaic *Homo sapiens* being transformed pretty quickly into behavioural moderns – is entirely artefactual. And even if this statement cannot be made for *Homo sapiens* in general, it can probably be made for the individual communities of early men that happened, wherever and whenever, to be baptised into symbolism.

So how did it happen? What triggered the symbolic revolution? It is worth looking at the question for the purposes of our inquiry, because of the close temporal connection (if no other connection) between the emergence of symbolic thought and the emergence of religion.

The short answer is that nobody knows. A rather longer answer is given in *The Selfless Gene*.[9] But one of the most respectable suggestions comes from the anthropologist David Lewis-Williams.

*Possible representation of the 'tunnel' phenomenon,
Ifran Takza, River Draa, Morocco.*

If we think about the cave paintings of Upper Palaeolithic Europe we think, rightly, of animals. They are beautifully done by artists who were superb naturalists. They knew how rhino ran; how deer threw their heads when they panicked; how the intestines of disembowelled bison uncoiled. But they are not simple bestiaries. Galloping amongst the naturalistic animals are gorgeous monstrosities: therianthropes (part-human, part-animal hybrids) and chimaeras assembled from the parts of different animals. The artists make use of natural features in the rock to give the animals life. A bulge in the rock becomes a head or the swelling of a muscle. There is an inescapable feeling that the visible animal is pushing through into the cave from a world on the other side of the stone wall. The animals are never depicted as running on the ground. Indeed, they never have any real spatial context other than by reference to the cave wall itself. You never see them against a mountain, or a tree, or fording a river. They seem to float. There are other things too: banks of wavy and zigzag lines, chequerboard patterns, ladders, webs, honeycombs and dots. These other images are often superimposed on the laboriously executed animal figures. Are they simple vandalistic graffiti painted by Cro-Magnon yobs? If they are, the yobs seem to have got bolder or busier as the Upper Palaeolithic wore on. By the time we get to the Magdalenian era (about 17,000–12,000 years ago) the patterns are everywhere. They are often in vertiginously or claustrophobically inaccessible places.

Paintings from this period are common in Africa too – indeed, they continued to be done there until the nineteenth century[10] – but they tend to be found on open, sheltered rock faces rather than in the depths of caves. In Africa, human figures are much more common: there are almost as many as there are animal figures. Human figures are very rare indeed in Europe.[11] The African humanoid figures, although again the work of master artists, are often bizarrely elongated. Often they are bent over at the waist, hold their arms behind their back, and some have unmistakably erect phalluses. Sometimes they are pierced with spears or arrows; sometimes some substance seems to be pouring out of their nostrils. Sometimes animals are being led with ropes.

Therianthropes and chimaeras are ubiquitous, and so are the geometric patterns.[12]

What is going on? There are four theories. The first is 'Don't know', and is always to be taken very seriously. The second is 'Art for art's sake', and is palpable nonsense. Many of the European paintings are painted in places that are immensely difficult and sometimes dangerous to get to. Perhaps the best-known example is the 'Shaft of the Dead Man' in Lascaux, south-west France. You crawl through a narrow crack and then have to be lowered five metres down a sheer drop to a ledge. Only then can you see the bird-headed man with four digits on his hand, about to be gored by a dying bison. It's no place for an art gallery, and it wasn't one. The 'pure art' idea can't explain, either, the existence or nature of the geometric patterns, let alone their juxtaposition with the images. It is contradicted, too, by the fact that many of the images are painted directly on top of earlier images. There is plenty of unused wall space.

The third theory, 'Hunting magic', used to be popular, but can explain the geometric patterns no better than the 'pure art' hypothesis. Nor can it deal with the dearth of missiles. If the idea of a painting was to put some sort of spell over a hunted animal in order to guarantee success in the hunt, one would expect to see many of the depicted animals transfixed with spears or arrows. But very few of them are – only 3–4 per cent. We also have a fair idea of the important prey species, and they are not the ones most commonly depicted on the cave walls.[13]

And so we are left with Lewis-Williams' ideas – ideas originally derided, but now the nearest thing to orthodoxy that the world of Upper Palaeolithic archaeology has. He may have tried to press things slightly too far (his ideas, although powerful, are not necessarily the key that unlocks everything in prehistory), but as an explanation for the cave paintings themselves Lewis-Williams is as good as we can get.

He had wondered what to make of the rock art of southern Africa. The scales fell off his eyes when he read interviews with San bushmen recorded in the 1870s by the German philologist Wilhelm Bleck and Bleck's sister-in-law Lucy Lloyd. The paintings and engravings, said the bushmen, were made by people 'full of supernatural power' – shamans.[14] The shamans went on their journeys after arduous foodless and drinkless trance dances that might last twenty-four hours. The small blood vessels inside the shaman's nose would often become fragile from the dehydration and burst – hence the nose bleeds shown in the images. The therianthropes depicted the shamans just as they transformed into the animal forms necessary for travel in the spirit-world, and many of the other images were what they had seen when they were there. The rock art images were illustrated travel books.

The geometric patterns are the ubiquitous entoptic phenomena associated with many altered states of consciousness. They are what you see when you enter, with the help of entheogens or otherwise, an altered state of consciousness. There is a fair chance that we will all see them when we are dying. Such altered states are commonly associated, too, with a feeling that the normal boundaries of the body have been changed – hence the elongated figures. If Andrew Newberg had been able to put one of the San shamans into his SPECT scanner during a spirit voyage, he would doubtless have found alteration in parietal lobe activity similar to that seen in those meditating nuns. The animals on ropes were 'rain animals' being captured by the shaman and dragged back to this world. When they were cut or milked, rain would come. The bent-over posture was the position adopted by the shamans in the trance dance, and as ecstatic consummation came, so did an erection.

The pierced figures are similarly illustrations of the shamanic process. Mircea Eliade[15] and Joan Halifax[16] compiled gruesome catalogues of the ordeals gone through by shamans – particularly

Negative imprints of human hands on the cave wall at Gargas, in the French Pyrenees. This is a strange place to have merely decorative images. They seem to be votive. Perhaps they depict hands touching the thin veil (represented by the cave wall), separating this world from the world to which the shamans travelled.

Upper Palaeolithic representations of chimaeric creatures. Cueva de los Casares, Guadalajara, Spain.

Floating, dream-like animals, superimposed on one another. Note the striated diagrams – possibly representations of the 'tunnel' through which the shaman passed to get to the world depicted, or maps of the other world. (Cueva de la Pasiega, Cantabria, Spain.)

in their initiation. To become a shaman is often no easy matter. In many cultures the apprentice shaman experiences, in his initiatory trance voyage to the spirit world, torture, death and dismemberment. His shattered body is then rebuilt and reborn into his terrestrial body. He will never be the same again: he now has – because the other world is his new birthplace – a right to be in the abode of the spirits, and an instinctual knowledge of its customs. He has dual citizenship, and can broker deals for his earthbound but spirit-affected clients.

The tortured shaman is everywhere. In the Australian bush an Aboriginal shaman's sides are cut open, his internal organs removed and a snake implanted in his head. Amongst the Pomo Indians of North America, the initiate endures the 'Ghost Ceremony', in which he is tortured, dies, and rises. Late Maya ceramics and Chinese Shang Dynasty cast bronzes both show how the apprentice gives himself up to the initiating spirit-cat (a jaguar in Central America, a tiger in China). The pose is almost identical in each: the cat's teeth penetrate the skull of the screaming shaman. An Inuit shaman from Arctic Canada drives his seal harpoon through his own abdomen.[17]

The pierced shaman is a common motif in many world religions. St Sebastian is said to have been pierced by many arrows but not to have been killed by them. Perhaps the story is an oblique reference to those pierced shamans who travel to other worlds and then return. (Church of St Sebastian, Via Appia, Rome)

We cannot stop now. We have to go on and note that St Sebastian, bristling with arrows, looks beatifically up to heaven, looking forward to his trip, and that St Teresa of Avila's encounter with an angel could be lifted straight from Bleck's transcripts:

I saw in the hands [of the angel] a long golden spear, and at the point of the iron there seemed to be a little fire. This, I thought, he thrust several times into my heart, and I thought that it penetrated to the entrails. When he drew out the spear he seemed to be drawing them with it . . . The pain was so great that it caused me to utter several moans.[18]

In every church there is either a model of a Jewish shaman nailed to a piece of wood on a rubbish dump outside Jerusalem, or a model of the gibbet on which he was strung.

A shaman whose piercing, journeying to another world and return to this one are historically attested. A reproduction of the crucifixion of Jesus, extrapolated from the findings on the Turin Shroud. (Santa Croce in Gerusalemme, Rome)

The very first real evidence of human symbolising and religion is, or comes at exactly the same time as, evidence of shamanic journeys. Those shamanic journeys may very well have been facilitated by hallucinogenic substances – perhaps the fly agaric mushroom, *Amanita muscaria*, in Europe. Many of the ways of inducing shamanic trance still used in Africa are arduous.[19] One can't blame shamans for wanting a chemical short cut.[20] But did the shamanic journeys themselves, or perhaps the hallucinogens used as chemical wings, enable those previously non-symbolising brains? Did they flick some sort of neural switch? Did they allow men to use in a symbolic way the awesome and more than adequate software that was inside their heads? Travel broadens the mind: did spirit travel transform it?

The entrance to the Neolithic tomb at Newgrange, County Meath, Ireland. The spiral decoration is thought by some to refer to the 'tunnel' experience of shamans travelling between this world and another – imagery invoked in the tomb because the dead would make the same journey.

The Lewis-Williams hypothesis is that it did.[21] He emphasises the role of hallucinogens, saying, in effect, that they flung open the doors of symbolic faculty. What clinches the case for him is the close association of sophisticated symbolism (the rock art) and the geometric patterns.[22] He contends that these patterns are repeated endlessly throughout the prehistorical world because the experiences that generated them were repeated endlessly. He sees Neolithic funerary architecture as a physical representation of the laminated shamanic cosmos. The Knowth passage tomb, for instance, one of the great monuments of Neolithic Ireland, has a corbelled roof above the main chamber, giving the impression of a tunnel spiralling upwards – a deliberate representation, he says, of the tunnel through which the shaman or the cardiac arrest patient passes en route to another realm.[23] Engraved into the wall of the Neolithic flint mine at Harrow Hill in England is the criss-cross pattern seen in Upper Palaeolithic cave art and reported by shamans and drug-takers.[24] '[I]f we look below the surface of [Irish and Near Eastern Neolithic architecture],' writes Lewis-Williams, 'we can detect similar neurologically generated building blocks.'[25]

One of the two passages in the Neolithic tomb at Knowth, County Meath, Ireland. Lewis-Williams and others believe that this, together with the corbelled roof-box, is a deliberate architectural reproduction of the shamanic 'tunnel' experience.

This is all tremendously exciting stuff, and many are tremendously excited by it. Nobody has fashioned any more promising instruments for probing inside the prehistoric mind. But the fact that we haven't got anything better doesn't mean that Lewis-Williams can tell us everything. Can we really go as far as Graham Hancock?

> Arguably art and religion are our two most prized cultural institutions, the saving human graces from which have arisen many of the most noble virtues and glorious achievements of our species. But if Lewis-Williams is right, we need to be brutally honest about where the first incarnations of these institutions came from. Their birth was not assisted by the operation of any of the faculties that we admire in the twenty first century – such as reason, intelligence, the scientific application of logic, sensitivity to nature, or even consciously driven creativity. Instead it seems that art and religion were bestowed upon us like secret and invisible powers, by inner mental realms that our societies now despise and legislate against – the realms of altered states of consciousness, most commonly entered (by us or by our predecessors) through the consumption of potent hallucinogenic drugs.[26]

British writer Graham Hancock, who proposes that the worlds reported by shamanic travellers are real.

This conclusion embodies two dubious equations. It suggests that once you've explained art you've explained all symbolism, and that once you've explained symbolism you've explained religion. Neither is necessarily right. But even if you've got full-blooded symbolism coursing through your psyche, none of the 'What if?' questions likely to intrude into an Upper Palaeolithic mind – including 'What happens if I die?' – begins to nudge us anywhere nearer religion. And even if you can explain the *fact* of religion, you're a million miles from explaining its contents.

This last point, to be fair, Hancock expressly acknowledges. Indeed, much of his fascinating book, *Supernatural*, is an attempt to expand Lewis-Williams' theory to account for why religions believe what they believe. It takes him to some strange places. Unable to deny the overwhelming sense of the reality of the visions he had on ayahuasca and DMT (Newberg might have helped him to deny it, of course), and struck by what he saw as parallels between experiences on hallucinogens, the accounts of UFO abductees and the stories in all the world's mythologies of visitations by fairies, elves and so on (often dancing in rings and then vanishing to other places, or being swept off in flashing sky vehicles), he concluded that alien abduction in the Upper Palaeolithic, together with marriages to 'spirit spouses' (producing hybrid man–alien children), might 'have provided the catalyst that . . . transformed anatomically modern but dull, uninspired and spiritually void humans into behaviourally modern, innovative, spiritually aware and interesting humans'.[27]

There are many variations on the thesis. Terence McKenna, for instance, sees the visions that one gets on psilocybin mushrooms as messages from alien intelligences. Mushrooms are ideal vehicles for the information that the aliens want to convey because their spores are so very robust and so easily transportable. They easily survive the icy rigours of interstellar space. '[I]f someone were designing a bioinformational package, a spore is how you would go. Millions of them pushed around by light pressure and gravitational dynamics would percolate throughout the galaxy.'[28] Mushrooms are a form of promiscuous mycological evangelism. The universe is being cunningly enseeded by billions of edible tracts.

For Hancock, it is neither here nor there that a shaman's body does not leave the campfire when he goes into a trance, or that an alien abductee's spouse confirms that the abductee was physically beside them when they report being in the spaceship, being reproductively manipulated, or that a subject in an official trial of DMT never leaves the hospital bed while wandering in another world. In each of these cases, says Hancock, there was a real journey by the real subject to a real place. The abductee stories are full of accounts of nurseries and strange, pale, hybrid children. The aliens want to breed with humans to improve alien stock in some way – perhaps so that the aliens will be able to materialise better than they apparently can. *Homo sapiens* has been the beneficiary of this interbreeding. Our ancestors have actually copulated with gods – for which read aliens. We have divine star-blood running in our veins. It is not surprising that the first infusions of that blood – or at least contact with the gods themselves – produced some sort of revolution. The divine spark that set alight human culture and religion was brought on a spaceship and implanted in the womb of human women. 'You have set eternity in the hearts of men,' say the Christians.[29] If Hancock is right, that is almost correct, but he would modify the assertion to: 'The "gods" have set the tendency and the ability to think about eternity in the uterus of women.'

For the more pedestrian of us – those whose ancestors were left behind when the fortunates were smuggled to other worlds – there were still the chemical vehicles. Magic mushrooms, peyote, ayahuasca and the many other compounds known to be used sacramentally enabled a 'democratisation' of altered states of consciousness. Anyone could jump, cheaply and relatively pain-lessly, on a bus to another universe.[30]

But even we unrealised ones have a visible genetic legacy of alien abduction – and it is not just our tendency to go to church. DNA-like helices appear in many hallucinogenic experiences. Sometimes snakes coil in double helices. This, thinks Hancock, is a real reference to DNA. The aliens have stored huge amounts of information on what we conventionally see as the genetic detritus of evolution – the so-called 'junk' DNA. This has no known function, and seems to consist of meaningless, repeti-tive sequences. It is not meaningless, says Hancock.[31] It is a repository of ancient and very important information, deposited there by the aliens who awoke our species and have been inter-ested in us ever since. Normally the information on that DNA is inaccessible to us, but on hallucinogenic drugs – or presum-ably in altered states of consciousness induced in other ways – we can in some ways peep into that mysterious library. It reveals some of the most fundamental truths about what we are and where we came from. It is no surprise that Francis Crick is alleged to have been under the influence of LSD when he had his Eureka moment and saw the structure of the DNA molecule.[32] The DNA was speaking directly to him about itself, and so about him.

Looked at through this lens, the history of religion is colourful indeed. When Moses spoke face to face with God at the top of Mount Sinai, there was 'thunder and lightning, as well as a thick cloud on the mountain, and a blast of a trumpet so loud that all the people who were in the camp trembled . . . Mount Sinai was wrapped in smoke, because the Lord had descended upon it in fire . . . the whole mountain shook violently.'[33] It is not surprising: a spaceship was hovering or landing. When Moses came down from the mountain, 'the skin of his face shone'[34] so that Aaron and the others were frightened.[35] Moses wore a

veil for much of the time after that.[36] Why? Perhaps radio-
activity from the spaceship had caused some radiation damage.[37]

Spacecraft appear regularly in the Bible and other ancient
accounts, if you believe Hancock.[38] Elijah's abduction into one
is actually witnessed: 'As [Elijah and Elisha] continued walking
and talking, a chariot of fire and horses of fire separated the
two of them, and Elijah ascended in a whirlwind into heaven.'[39]

What is the first vision of Ezekiel but a UFO, populated by
a crew of four alien beings? He sees it approach and land. (The
bracketed interpolations are mine):

> As I looked, a stormy wind came out of the north: a great cloud
> with brightness around it and fire flashing forth continually, and
> in the middle of the fire, something like gleaming amber. [The
> cloud is perhaps a vapour trail. The flashing fire might be from
> some sort of propulsion device, or perhaps landing lights. The
> central component, where the crew are, is the illuminated
> cockpit.] In the middle of it was something like four living crea-
> tures. This was their appearance: they were of human form. [All
> aliens in the stories are.] Each had four faces [perhaps a helmet
> with, as we see later, various devices on each side which gave
> each face the appearance of animals?] and each of them had
> four wings [which might be portable breathing tanks or, indeed,
> mechanical wings]. Their legs were straight, and the soles of
> their feet were like the sole of a calf's foot [a reasonable descrip-
> tion of an astronautical boot] and they sparkled like burnished
> bronze [a very wise precaution: a reflective suit would reflect
> away incident heat and make homeostasis easier] . . . I saw a
> wheel on the earth beside the living creatures, one for each of
> the four of them. As for the appearance of the wheels and their
> construction: their appearance was like the gleaming of beryl;
> and the four had the same form, their construction being some-
> thing like a wheel within a wheel. When they moved, they moved
> in any of the four directions without veering as they moved. [It

is some sort of sophisticated, shiny buggy.] . . . Over the heads
of the living creatures there was something like a dome, shining
like crystal, spread out above their heads [the roof of the
cockpit].[40]

Modern UFOs might be more slick and efficient than the version
described by Ezekiel, but then they have had two and a half
thousand years to improve the design. The winged discs of Sumer
and ancient Egypt, the flying carpets of Arabia, the sky-rafts
piloted by ancient Chinese star-spirits, the Vimana aircraft
mentioned in the Hindu texts and possibly even the broom-
sticks of the rather backward European witches are, in Hancock's
world, different models of UFO.

What about the alien gods themselves? The Bible gives an
explicit account of the impregnation of human women by the
aliens, according to Hancock:

> When people began to multiply on the face of the ground, and
> daughters were born to them, the sons of God saw that they
> were fair; and they took wives for themselves of all that they
> chose. Then the Lord said, 'My spirit shall not abide in mortals
> for ever, for they are flesh; their days shall be one hundred and
> twenty years.' The Nephilim were on the earth in those days –
> and also afterwards – when the sons of God went in to the
> daughters of humans, who bore children to them. These were
> the heroes that were of old, warriors of renown.[41]

The vision of Ezekiel, according to Benny Shanon of the Hebrew
University, Jerusalem,[42] sounds just like the sort of hallucination you see if you are on ayahuasca. Other core Jewish events,
such as Moses' encounter at the Burning Bush, have similar
explanations. Ayahuasca isn't found in the Near East, but cocktails of acacia and Syrian rue, which certainly are, produce a
very similar effect. Perhaps John, sitting on Patmos, had taken

a dose of something like that. What else could have caused him to be 'in the spirit on the Lord's day'?[43] And isn't there something creepily familiar (and alien) about what he sees?

> [I]n the midst of the lampstands I saw one like the Son of Man, clothed with a long robe and with a golden sash across his chest. His head and his hair were white as white wool, white as snow; his eyes were like a flame of fire, his feet were like burnished bronze, refined in a furnace, and his voice was like the sound of many waters. In his right hand he held seven stars, and from his mouth came a sharp, two-edged sword, and his face was like the sun shining with full force. When I saw him, and fell at his feet as though dead. . .[44]

And so it goes on. It is all good fun, but soon gets rather predictable. In Mormonism the angel Moroni who hands down the text of the Book of Mormon is actually an alien. When he was twelve years old Mani, the founder of Manichaeism, was visited by an angel (an alien, of course, or possibly a temporal lobe seizure) who visited him regularly throughout his life, accompanied always by a flash of lightning (spaceship/seizure: it doesn't much matter which, since both take you to another world). St Paul was involuntarily initiated as a shaman on the road to Damascus (again, either epileptically, by alien apparition or by abduction: they are much of a methodological muchness). When Hildegard of Bingen, illustrating the twelfth-century Rupertsburg Codex, painted a child quickening in the womb, attached umbilically to a quadrangular device in the sky above it, she was acknowledging that aliens gave life, or at least infusions of knowledge, to humans. For slow readers, the plate rams home the esoteric point: just next to the child are a mushroom and a goblin. The initial vision of Bernadette of Lourdes occurred when she was exhausted and had just forded an icy river – ideal circumstances for an altered state of consciousness. She sees what she describes as a 'Lady' (who only later mutates into the Virgin) who tells Bernadette to eat a 'herb' – no doubt a psychoactive one. In her later, frequent trance states, Bernadette communicates with 'the Lady' in just the same way as abductees communicate with their contacts on the mother-ship. Joan of Arc seems to have been similarly uncertain that she had seen the Virgin. In her testimony to her inquisitors, she spoke interchangeably about 'white ladies', 'good ladies' and the Virgin. And she also told them all about a 'fairy tree' – a tree that Hancock invites us to infer had psychedelic properties.

And then there is Jesus:

> After its foundation around two thousand years ago, [Christianity] was, at first, an overtly shamanistic religion. This is hardly surprising, since Christ was so obviously and so profoundly a shaman. It is not only his pedigree as a half-human, half-divine hybrid that makes him so, or his heaven-sent gifts as a healer. His ordeal of crucifixion and piercing followed by death and subsequent resurrection as a spiritualised being equipped with the power to save souls is essentially the story of the wounded man – the story that is told by all shamans everywhere of their own initiatory agonies, death and resurrection.[45]

Hancock goes on to reiterate the notion that the Gnostics were in some way the original Christians – the true inheritors of Jesus' teaching; the only ones who really understood what he was talking about. This is unhistorical nonsense, in which Hancock joins the unhappy company of Dan Brown and much of the New Age movement. But the rest of his thesis is brave and internally coherent.

But I don't believe a word of it. There are many, many reasons why not. There are two that are particularly important for present purposes.

First: cherry picking. It's not fair to rest a whole thesis on Mount Sinai (which anyway sounds much more volcanic than extraterrestrial), on Elijah (if the vehicle was a saucer, you might have expected Elisha to do better than the rather lame 'chariot'; also Elijah never came back – almost all recorded abductees, at least in modern times, do), on Ezekiel (the only really mechanical-sounding passage), on the strange Nephilim (who baffle everyone), on Paul (who might have been epileptic, but so what?) and on John of Patmos (who no doubt was in some sort of religious trance – and what of it?). There are many more encounters between God and mankind, even in the Judaeo-Christian tradition alone, and most of them aren't remotely like that. If you're purporting to explain away religion, you need to explain away the general pattern, not the weirdnesses that happen to suit your case.

And second it doesn't begin to explain the *content* of religion. Let's suppose, for the sake of argument, that Hancock is right to press the Lewis-Williams hypothesis as far as he does. Let's suppose that aliens really did come down sometime in the Upper Palaeolithic, appear somehow to our ancestors, and somehow bequeath them and us an artistic impulse, a general symbolising facility and a religious tendency. Let's suppose, too, that Mount Sinai and all the other supposed examples of alien intervention are indeed that. It is strange, to start with, that these aliens should bother to hand down the Ten Commandments. Did they do so because they cared about the welfare of the creatures whose DNA they had mixed with theirs; whose cells they had chosen to use as filing cabinets? But if they did bother to draft and promulgate the rules, it is even stranger that, in their other apparitions to mankind, they should have given different drafts. The world's religions aren't the same. The Ten Commandments and the Book of Mormon (to take two examples of texts supposedly handed down from on high) aren't expressions of that elusive perennial philosophy; nor do they agree with one another. Are there competing sects of aliens out there, engaged for the last forty thousand years in an evangelistic turf war for the souls of the earth-dwellers? Were the spaceships full of predatory preachers? Come on.

I have no basic problem with the idea that an experience of something very strange might have triggered religion. I would go to the barricades to contend for the idea that the content of religion is not something that we are likely to have made up for ourselves. But even if Lewis-Williams can explain art (and all he can really give us is a close temporal association between the first art and the experience of altered states of consciousness), he can't give us religion. Nor can Graham Hancock. Nobody knows what triggered other elements of our symbolic facility – language, for instance. But even if we knew the source of the first, epoch-making 'What if?', and even if it bears an implicit capacity for conceiving worlds other than our own, we would still have an explanation only for religion *per se* – not for the religions we actually see or which we know have existed. A content-less religion really isn't very interesting or significant.

Newberg contends that faith is essential to allow us to trust the beliefs that are generated by our biochemistry. If we didn't trust them, he says, we'd be paralysed by our doubts in an adaptively disastrous way.[46,47] But 'faith' by itself is a completely meaningless idea. If someone tells you that he has 'faith', you're none the wiser. Faith in what? Nothing that might be generated inside our skulls can sensibly be suggested to have anything other than a deleterious adaptive effect. The floundering of Richard Dawkins and the memeticists illustrates that natural selection can account satisfactorily for neither the existence nor the content of religion.[48] To say that the architecture of the human mind is an architecture that will tend to favour religion is true and important, but begs the question, 'Why is it like that?' There's no evidence that it was formed by hallucinogens or sculpted by the crew of alien spaceships. For the moment, the best answer is the usual best answer: we just don't know.

Nonetheless, Hancock is onto something. There is a surprising overlap between the images seen by very different people on very different drugs. Snakes abound. So do therianthropes. A Brazilian shaman on ayahuasca will see a human with a jaguar's head. The neurophysiologist Kluver, taking mescaline, saw a human head sprouting the fur of a cat and finally mutating entirely into a cat's head.[49] Another European subject, dosed with hashish, 'thought of a fox, and instantly I was transformed into that animal. I could distinctively feel myself a fox, could see my long ears and long bushy tail, and by a sort of introversion felt that my complete anatomy was that of a fox.'[50]

On my desk stands an Inuit *tupilaq*, carved out of a lump of narwhal tusk. Its body is upright, like a man's, and it has humanoid legs and hands. But its head is like a seal possessed, with wildly flaring nostrils, pitiless black eyes, a lantern jaw and perfect human teeth. Out of the back of its head sprouts a small, surprised polar bear, and between its legs are flat, fierce, generically malevolent faces of the sort seen everywhere that men dream.

The temples of many religions groan with exotic therianthropes: the jackal-headed Anubis, the ibis-headed Thoth and many other examples from ancient Egypt; elephant-headed Ganesh in Hinduism; the mysterious Jaguar–Baby of Mayan religion, and so on.

*Animal figures from La Pasiega, Spain. Although beautifully
executed, the figures are not oriented in relation to each other in a
naturalistic way: they seem to float. Also note that some parts of
the animals are deliberately missing – as if they literally have a
foot in another world, just beyond the cave wall.*

*Floating 'spirit animals', hands passing against the cave wall (possibly
seen as a membrane separating this world from another), a presiding
human figure one of whose arms looks as if it is transforming into a
snake, and a possible representation of the 'tunnel'. 9,500–13,000
years old. Rio Pinturas, Santa Cruz Province, Argentina.*

Therianthropes: The Las Limas sculpture from the Mexican state of Veracruz, in the Olmec heartland. A youth holds a human-jaguar hybrid. (Xalapa Anthropological Museum, Veracruz)

Therianthropes: The hawk-headed Egyptian god, Horus and the jackal-headed Egyptian god, Anubis.

Hancock thinks that this extraordinary commonality, together with the overwhelming sense of reality that he got when he wandered in his entheogenic world, means that the figures are real, and that religion is a celebration of those figures and the world in which they live, and a means of getting to that world. There is a sense in which I agree with him. The therianthropes (for instance) are mythical Jungian archetypes: part of the heritage bequeathed to us; real, mythologised memories embedded in our neurones, accessed and vivified by entheogens, exhaustion or extreme asceticism. The old memories are now so encrusted with myth that it is difficult or impossible to see their original outlines. As a means of telling us about ourselves and where we came from, they are no doubt potentially invaluable, but they are encrypted in long-lost languages. There is no reason to suppose that they

Therianthropes: The elephant-headed Hindu god, Ganesh.

tell us anything at all about the origins of religion, let alone that they tell us everything.[51]

The origin of religion remains a baffling conundrum. Plants and shamanistic asceticism don't help. Religion isn't a plausible by-product of the symbolic revolution. It flouts the most basic axioms of natural selection: Darwinism would have strangled it at birth.[52] Memes, if they are to be taken seriously at all, explain only dissemination, not origins. So what's left? The most likely candidate is a religious experience of a profound kind. It would need to be far more intimate and self-explanatory than the schizoid shards of experience given by peyote.

Therianthropes: Christian therianthropes: A saint preaches to dog-headed men and an ikon of the dog-headed saint, St Christopher

CHAPTER 12

Angels or Demons?
The Suppression of Religious Experience

Reports of the prohibition of shamanism and of its persecution and suppression by the Lamaist missionaries run like a red thread through the history of the second conversion of the Mongols from the 16th century onwards. Anti-shamanist resolutions such as the legislation of Dzasakhtu Khan (1558–82/3) and the edicts of Altan Khan of the Tumet (1557/8) which forbid under threat of punishment the possession of Ongghot figures and their worship through bloody sacrifices, form the prelude. The stimulus for them came from the Third Dalai Lama himself. The shamanist idols were to be replaced by representations of the Buddha and, especially, by such 'terrifying' deities as the six-armed Mahakala.

(Walter Heissig, *The Lamaist Suppression of Shamanism*, trans. G. Samuel, 1970)[1]

By 1965, Timothy Leary, formerly a clinical psychologist at Harvard, was recognised as the main advocate of the use of hallucinogenic drugs. His notoriety had increased after his dismissal from Harvard in 1963, his drug-fuelled parties on the famous Millbrook estate, and his 1964 publication of *The Psychedelic Experience*.

In December 1965, while she was crossing to the United States from Mexico, Leary's daughter was arrested. She was carrying a small amount of marijuana. Leary took responsibility, and was duly convicted and sentenced to thirty years' imprisonment, a $30,000 fine, and ordered to undergo a course of psychiatric treatment. His conviction was soon quashed,[2] but in December 1968, Leary was again arrested. The court found that he was in possession of two roaches of marijuana. He was sentenced to ten years' imprisonment. While in custody he was convicted again in relation to another marijuana offence committed in 1965, and given another ten-year sentence. For possessing less than half an ounce of marijuana, the sentence was thus twenty years.

Since he had designed many of the psychological tests to which he was now subjected, he was able to paint a picture of himself as a rather grey, conventional conformist, fascinated by gardening. He was duly placed on gardening duties in a low-security prison, from which, without any difficulty, he escaped in 1970. His wanderings make a great story. He found his way eventually to Afghanistan, where US officials, acting with distinctly dubious legality, arrested him in 1973. The US officially saw him as public enemy number one. 'He is the most dangerous man in America,' growled Richard Nixon. Back in America, Leary now faced ninety-five years in prison. He was in fact released early in 1976.

The point about all this is that, whatever you think about the control of drugs and the cocky narcissism of Timothy Leary, the US response was ridiculously, laughably disproportionate.[3] Why were Leary and the drug culture he personified persecuted with such hysterical zeal?

When Constantine became emperor, the old mystery religions, some of which no doubt used psychoactive substances, were, after an initial period of toleration, suppressed or pressed into hiding. When Catholic missionaries arrived in South America, they lost no time in trying to stamp out the magic mushroom cults they found there. Modern Protestants continue the work that the Catholics began.[4] Even the gently psychoactive plant kava, beloved of Pacific Islanders, was violently disapproved of by the bow-tied missionaries who landed in search of souls.

It is not just hallucinogenic substances that have incurred wrath. Shamanism generally has been hunted almost to extinction, and not just by Bible- and musket-wielding Conquistadors and the prosecutors of the European witch trials. Even the normally accommodating Buddhists of Mongolia have been aggressively intolerant of shamanic practice there. If things go on as they are, it won't be long before there is no one left who can be in four places at once, fly as fast as a bullet, or make rain. You might have thought that Michael Persinger, starved of funds and having to pay for all his research out of his own pocket, would have enough to contend with without having evangelical Christian groups picketing his laboratory, contending that both he and his 'God helmet' are demonic, but they are regularly there.

John Mack, Professor of Psychiatry at Harvard Medical School, had the effrontery to ask some courageous and fundamental questions about the experiences of the alleged 'alien abductees' he had interviewed.[5] It very nearly cost him his job. It did cost him his reputation. The pattern has been depressingly repeated across university campuses worldwide. Academics beware: asking the big questions can seriously imperil your tenure. There's a nasty kind of censorship going on in the world of scientific publishing too. Karl Jansen has found it impossible to publish in mainstream journals his groundbreaking work on the relationship between ketamine and NDEs.[6]

But why? It is, of course, rather misleading to lump drug control in with witchcraft, evangelism and academic freedom. Different issues apply to each. But the conspiracy theorists have no difficulty with the conflation. They say that what is happening here is what has always happened: humans are being denied their birthright – being denied access to the very experiences that might have kindled their quintessential humanness. This is Prometheus all over again, they scream.

Prometheus, who has stolen fire back from the gods, is punished by Zeus by being chained to a rock and having his liver eaten by eagles. The liver regrows so that the torment can be repeated each day. A picture of general authoritarian disapproval of human presumption?

Zeus had petulantly withheld fire from the world. Prometheus stole it back, lighting and warming mankind again. Zeus, furious, chained Prometheus to a rock, and eagles ate his liver every day. The liver cruelly regrew, so that his torment would never end. If humans are allowed glimpses of the other worlds of which they are properly citizens, the rhetoric goes, the scales will fall off their eyes. They will want to be free, and the denizens of those other worlds will give them the power to be free. The Zeusian forces of oppression are terrified at the thought of this turbulent freedom, and accordingly do their best to bolt and double-bolt the portals.[7]

I'm not as dismissive of this Promethean explanation as I might sound. The distinctive note in many of the justifications for suppression is simple, straightforward fear. But fear of what?

234

Often it is fear of the cherished status quo being upset. Often it is fear of uncertainty: legislators of all sorts like more than anything else to know where they stand. They will legislate against imponderables and uncertainties for no better reason than that they are imponderable and uncertain.

The Christians have often been offended by the apparently close and apparently blasphemous parallels between the sacrament of the hallucinogen and the sacrament of the Eucharist, and their reactions to psilocybin mushrooms in South and Central America, iboga in Africa and kava in the Pacific Islands are hardly different in nature or degree to the arrogant way in which they have dealt with any local opposition to Christianity wherever they have encountered it.

Often, no doubt, the very democratic nature of a psychoactive sacrament that works for all has been seen as an evil, revolutionary challenge to the priestly prerogative. 'The advantage of the mushroom', wrote R. Gordon Wasson, 'is that it puts many (if not everyone) within reach of [the ecstatic state of the great mystics] without having to suffer the mortifications of Blake and St John.'[8] Priests like being the only ones able to peek behind the veil into the Holy of Holies. If you tear down the veil chemically and let all the hoi polloi in, there will be outrage from the established religions, prophesied Aldous Huxley. 'Instead of being rare, premystical and mystical experiences will become common. What was once the spiritual privilege of the few will be made available to the many. For the ministers of the world's organized religions, this will raise a number of unprecedented problems.'[9] No longer will religion be a matter of symbols and a response to those symbols. If they've had the real thing, the congregation isn't going to be easily fobbed off with a set of symbols. 'The perusal of a page from even the most beautifully written cookbook is no substitute for the eating of dinner. We are exhorted to "*taste* and see that the Lord is good".'[10]

This is really just a special case of the general Promethean worry.

But there is a more fundamental reason for the Judaeo-Christian suspicion of exotic mystical experiences, whether generated pharmacologically or otherwise. It is expressed in the terribly clear and violent injunctions against witchcraft and contacting the dead.[11] The injunctions are there not because those things are nonsense, but precisely because they are not. More generally, there is a clear insistence that some things are out of bounds, and that is not because Yahweh wants to keep them jealously to himself. He is no petulant Zeus, upset that the mortals are coming a bit too near to his own patch and worried that they might steal some of his glory.

It is perfectly true that the Genesis story of the plucking of the fruit of the knowledge of good and evil looks at first to be just what I am saying it is not: it looks like yet another of the Promethean stories of which the mythological books are full.

After the humans have eaten the forbidden fruit, 'the eyes of both were opened'.[12] God is displeased: 'Then the Lord God said: "See, the man has become like one of us, knowing good and evil; and now, he might reach out his hand and take also from the tree of life, and eat, and live forever."'[13] It is that worry that led to the expulsion from the Garden.[14]

I have dived elsewhere as far down as I can get into the meaning of this passage.[15] It is undoubtedly a passage of immense difficulty and subtlety. But the nuances don't matter for present purposes. All we have to do for now is to note that this story is right at the beginning of the Bible, and that much of the rest of the Bible is about the bloody, immensely costly consequences of that plucked fruit, the bloody, immensely costly attempts to mitigate those consequences and, finally, the bloody, immensely costly cure. Yahweh sends no eagle to rip the liver of mankind. There is an eagle, but he is Yahweh's deadly foe. Yahweh cautions as sternly as he can about journeys towards the eyrie, just as a mother snatches her child away from the fire.

The Judaeo-Christian tradition has been the victim of some dismal exegesis. Chanting, meditation and many forms of ecstatic practice are by no stretch of the most elastic expository imagination witchcraft or mediumship. Those biblical 'out of bounds' signs are no mandate for squeezing your liturgy, your devotion or your life generally into a sober, tightly buttoned Sunday suit. The biblical default position for praying is with your hands in the air. You should barely be able to see your hands through the incense – and incense, note, is a notoriously powerful olfactory driver, drenching the amygdala with neurotransmitters and likely to waft you upward with it. The Judaeo-Christian tradition urges us to dance like David before the Lord. And that, as we know, is likely to lead to an altered state of consciousness.

CHAPTER 13

Breathing God

Now I know in part; then I shall know fully, even as I am fully known.

(1 Corinthians 13:12)

There's more to religion than experience. An account of religion that doesn't mention morality or altruism is like an account of the Great War that doesn't mention Germany. 'Perhaps mystical visions are in fact nothing more than a bit of squelchy feedback in the temporal lobes,' writes Jack Hitt. 'But that's such a preposterously small part of what most people think of when they think of God, it seems insanely grandiose to suggest that anyone has explained away "God".'[1] But is your religion really religious if it does nothing to your mood; if it doesn't even hint at the ecstatic; if it doesn't sound a high note of longing that shows that you've entered into real citizenship of a supra-real country, or throb like a low jungle drum in time with your heart, setting the pace for your thought?

All but one of the great religions seem to have started with an experience. Think of Abraham's conversations with God, or Moses' epiphany at the Burning Bush; the Prophet Mohammed, alone in the desert, having the Qu'ran dictated to him by Gabriel; the Buddha's enlightenment under the bodhi tree; the Upanishads squeezed into the world after an agonising ascetic labour. If Lewis-Williams is right, the whole religious impulse itself was generated by the experience of an altered state of consciousness.

There are two odd things about the experiences described in this book. The first is their sameness. If you feed ayahuasca to a Bonn actuary, he will see the same snakes as an Amazonian parrot-hunter. Some of the Upanishads are Teresa of Avila on curry. St John of the Cross would have made an unexceptional Sufi. A stockbroker being prayed for in a South Kensington church shows the same rapid eye movement and lacrimation as an addict undergoing hypnosis in an attempt to break a heroin habit. Cut off the blood supply to an Inuit's brain, and there is a fair chance that he will wander down the same tunnel, towards the same light, as a profoundly hypoxic Pope. There are differences, of course, but they are not hugely significant. The Inuit will be met by his ancestors and his tribal gods. The Pope will be met by St Peter, or possibly Jesus himself, and might go through applauding ranks of cardinals. If we ever hear their stories, we hear that they wanted to stay in the light, but were sent back to complete some kind of mission.

The second odd thing is that these experiences often have some sort of moral corollary. People who have near-death experiences tend to say that they have made them better people; that they live more selfless lives; that they love more and hate less. The subjects in the Good Friday experiments, whatever the Federal Drugs enforcers say, seem to have been chemically catapulted into a fairly conventional niceness. Evangelical Christians who have a charismatic 'filling with the Holy Spirit' talk about their new-found love for the world and a strange new ability to forgive. Even Christina the Astonishing, lifted epileptically to the heavens, shrieked her way around medieval Belgium with the eternal destinies of her hearers at heart.

If experiences don't have this corollary, we should discount them as bogus, or reject them as worthless or worse. The American mystic Ken Wilber tells us unblushingly that he is one of a very small number ever to live in a permanent state of non-duality. He maintains it even during deep sleep, which is vanishingly unusual, even amongst the greatest yogis. 'Yesterday I sat in a shopping mall for hours watching people pass by,' he wrote in his book *One Taste*, a series of excerpts from his journal. '[T]hey were all as precious as green emeralds. The occasional joy in their voices, but more often the pain in their faces, the sadness in their eyes, the burdensome slowness of their paces – I registered none of that. I saw only the glory of green emeralds, and radiant Buddhas walking everywhere.'[2] If that's what you see from the summit of mystical experience, let's stay away. If someone tells me that he sees everything, but can't see pain, I doubt him. A mysticism blind to pain is blind. And let's not be intoxicated with the exotic language of transcendence. If I'm transcendentally immune to the suffering of others, my transcendence is evil. No decent person will make any peace at all with whatever causes a child to suffer, even if the purported broker comes straight from a monastery in the Himalayan fastness, fully accredited by all the elders of Nirvana. Obscene self-indulgence is obscene self-indulgence, however excruciating the ascetic trials through which it has gone. By their fruits, in short, you shall know them.

We know of only two states characterised by an absence of experience: unconsciousness and some forms of profound, anhedonic depression. In depression the inability to feel anything is itself an ironic type (and surely a particularly malignant type) of pain. When extreme Christian conservatives urge their followers to a suspicion of any sort of experience, they are really exhorting them to depression. The depression is often, and tragically, successful.

I expect that for most of us real depression means de facto faithlessness. We need to be able to *feel* that God is there in order to believe it. When feeling goes, the propositions in a creed just mock us with memories of what we once had. Perhaps the dark night of the soul is, or at least correlates with, clinical depression. But to find him in the heart of the darkness is the work of the really great saints.[3] I will never be able to do it, and I think few will. This book has looked at some of the mountain-top ecstasies of the saints, and rejoiced that some elements of those ecstasies are accessible to us. But perhaps the desert ecstasies are greater. And those I cannot write of. I devoutly hope never to taste them.

If certain places do things to my brain that I call 'religious', does it demystify, debunk or dereligionise the experience if it can be shown (which at the moment it can't) that the medium of the experience is, say, an electromagnetic field? I don't begin to see why. If our species has been here a while, as it has, one might expect our brains (and our souls, if they exist) to learn something of the language in which the universe speaks. To say that an electromagnetic field *is* God (and accordingly that the traditional God of the Judaeo-Christian tradition doesn't exist) is like picking up a French grammar book and saying, 'Look: the medium by which "Flaubert" communicates his message to us is now clear. We can therefore conclude that Flaubert doesn't exist and has nothing to say to us.' If I experience awe by my amygdala being tickled, that tells me nothing about the tickler except that he knows enough about me to tickle in the right place.

Jews and Christians are generally rather bad at getting out of their heads on anything and saying, 'I can see God more clearly now.' They have generally thought that they have been put in their heads for a very good reason and that, by and large, they will see things better from that vantage point. Yes, there is a fecund mystical tradition in both faiths, but, with a few spectacular and important exceptions, the mysticism has been an ecstatic appreciation of the immanence of God rather than of the transcendence of the self. Judaism and Christianity don't need peyote or ayahuasca. Red wine, wild but faithful sex, wild abstinence, promiscuous charity, mountainous surf and some strictly non-pantheistic tree-hugging will do the job far better.

I have few convictions, and only one unshakeable conviction. The one unshakeable conviction is that to live as a human being is itself, and inescapably, a religious act. To breathe consciously is necessarily a religious experience. To breathe more consciously, as the Buddhists splendidly know, is a more religious experience. Some things help us to be more ourselves, and therefore more holy. Other people help more than anything, but hills and meditation help too. For Geoffrey Household, it was wine that 'heighten[ed] human sympathy and human perception' and carried him, 'even if in times past with too much noise and extravagance of spirit, towards the possible man at whom I was intended to aim'.[4] Others would struggle without Bach, centring prayer, the music of mathematics, the shrine at Walsingham, or steak and kidney pie. To know someone, watch him dancing. And, perhaps paradoxically and perhaps not, the more formal the dance, the more impossible it is to conceal oneself from either the audience or oneself.

And now, perhaps, we can begin to make some sense of those strange eucharistic words, 'Take, eat: this is my body,' 'This is my blood you drink.' The Last Supper was an ordinary meal of ordinary things. It involved only a mildly psychoactive, everyday drug, alcohol, which Jesus expected would be at every table. The meal would inevitably be repeated whenever friends came together. And there's the mystery, hidden in the heart of the everyday. Being a wine-drinking, bread-chewing human is incurably sacramental. The basic stuff that keeps you alive keeps you holy. Air and water are sacramental, and indeed psychoactive. No wonder the Gnostics loathed the real Palestinian Jesus, striding through the Palestinian dust and sitting in wine-houses with hookers.

We have to note one other thing about the religious experiences we have surveyed. That is that they are all *relational*. Sometimes the relationship they express and expound is a relationship with the self: 'I feel happier with myself'; 'I am more at peace with myself'. Often it is a relationship with the worshipping congregation: 'I felt caught up in the exhilarating community of my fellow churchgoers/mosque attenders/football fans.' Sometimes it is with God himself, perceived as God himself – and often fragrantly and passionately erotic language is used to describe that sort of relationship. Even when the experience approaches that of Absolute Unitary Being – the ultimate goal of Eastern religion – the language is that of relationship: 'The boundaries between me and the universe were dissolved. I was finally what I was meant to be.' What an irony. You are most yourself when you are least yourself.

And yet every lover knows it. Self is only fully consummated when it gives itself away. By doing so it becomes all the more truly itself. When, in the throes of yogic climax or sexual union, the parietal lobes are joyously underperfused, melting the frontiers of the self, the self becomes not smaller, but larger and more quintessentially, uniquely, itself. 'The golden apple of selfhood, thrown among the false gods, became an apple of discord because they scrambled for it,' wrote C. S. Lewis.

They did not know the first rule of the holy game, which is that every player must by all means touch the ball and then immediately pass it on. To be found with it in your hands is a fault: to cling to it, death. But when it flies to and fro among the players, too swift for eye to follow, and the great master Himself leads the revelry, giving Himself eternally to His creatures in the generation, and back to Himself in the sacrifice, of the Word, then indeed the eternal dance 'makes heaven drowsy with the harmony'.[5]

The Eastern language of transcending the self is more helpful (since less open to ascetic and other abuse) than the Christian language of crucifying the self to death. St Paul was sometimes too potent a poet for the good of us weaker brethren. When something is transcended, it doesn't disappear. The language of transfiguration is even more accurate, and tallies even more neatly with the tales that come back to us from the Himalayan eyries, the monks' cells and the better class of opium den.

Susan Blackmore talks about one person whom she believes might be truly 'enlightened' in the sense meant by the mystics of the East. He lives, in other words, self-transcendentally, non-dually, in the state of Absolute Unitary Being. He is the Australian psychologist John Wren-Lewis. And his testimony is, 'Oh, all this, it's just the universe John Wren-Lewis-ing.'[6] Many Buddhists and Hindus would be outraged by this, and I don't know how accurate it is, but surely it is a description of a state of spiritual affairs happier than that described by the Christians who love Paul's language of the crucified self. Isn't it a lyrically philosophical picture of what must be meant by heaven rejoicing at the salvation of a human being? It's comfortably consonant with the idea of a personal God, wildly in love with you, just as you are, who hugs you like a son, and does everything possible to avoid your ablation.

Yesterday I went to a church where somebody (I'll call him John) sang, 'This is my pilgrimage; to be absorbed in you.' That's the antithesis of what the Christians, at any rate, have historically believed. They have always thought that John, Christianised, will be far more colourfully, unmistakably, hugely, solidly, magnificently, eternally John than ever before. When you're baptised, the water goes into you and makes you swell and coruscate with holiness and reality. You don't dissolve into it. And that, despite the sometimes clumsy attempts by the mystics to indicate how those inactive parietal lobes are making them feel, is the gist of those front-line dispatches from the wild frontiers of selfhood.

The one great religion that did not start with an obvious experience is Christianity. It started and will end with a person – the Great Shaman – who, if the Christians are right, did an entirely unmetaphorical journey to another world which wasn't remotely figurative, and returned, having done battle with the forces there, to fry fish on a beach. The Great Shaman said (with repulsive arrogance or psychotic delusion if it wasn't true) that he himself was the way to the experience everyone was looking for, and the experience itself. The body he returned in was a supra-physical body, a body built for sensation, for experiencing the very real sensualities of a world like ours, but even more like our world than our world is. It was a body no doubt crammed with receptors which would make our most orgiastic sensations on ketamine or LSD seem like filling in a tax return. When that body breathed, a single breath set running in those resurrected neurones impulses that out-sensationalised all the cumulative Dionysiac joys of the aeons. Bacchus fell prostrate, and kissed its feet.

A riddle:

It was a Tuesday night in an alehouse in York. In one corner was a cluster of musicians. On their fiddles and squeeze-boxes they played tunes rooted in the fields, woods and burial barrows of England. When their arms and fingers got tired, they reached to the table for more beer. The beer dissolved the boundaries between them and the others, and between them and the tune. The music only worked at all if the musicians flowed into each other, just as their notes flowed into each other's ears. A solitary prima donna would wreck the whole thing. The music could be played only by a single organism, but the absence of any one player would be fatal too. As the alcohol perfused the musicians' bodies, less and less blood perfused their parietal lobes.

They all stamped their feet, and in Yorkshire, just as in the Kalahari, rhythmic driving can do strange things to the brain. As my fingers flew over the holes of the tin whistle, too fast for thought, something other than thought reared out of my head and melted into the cloud of collective consciousness that hung over the tables. The Roman soldiers under the floor knew some of the cadences: they had heard them when patrolling at night, played on the bone flutes of the Brigantes. If the musicians were one with the cadence, they were talking with the Roman dead.

At the bar the chatter had stopped. Football and imminent redundancy were forgotten. Feet tapped, heads nodded, and there were the frowns of mystified recognition.

Into the pub came a brave and good man. He beamed at everyone through thick glasses. He walked round the bar, respectfully handing out tracts published in South Carolina. 'An invitation to live,' they said. 'I have come that you might have life,' Jesus told us, 'and have it in all its fullness.' He went over to the musicians' tables, gently putting a copy down in front of each of us. He knocked over someone's pint, apologetically mopped it up with a handkerchief that he took out of his dufflecoat pocket, and bought a replacement. But the music had stopped. He was slung out of the pub.

'Extraordinary,' said Ken. 'What an idiot.'

I was quiet.

Epilogue

For a day and a hot night I sat under a tree in Kerala. I heard the waves smashing into the sand, and did not bother to brush off the ants that swarmed over me. I drank water and watched my breath, fascinated. I let my legs go to sleep, and after a while they became part of the ground. Slowly I followed them. I went into the sea, and the sea went into me. Once every few hours I opened my eyes. When I did, the swaying palms were as familiar as my own hands. My mother had sung nursery songs to us both. Often an Indian crow was close by, its head cocked to one side, watching me. I warned myself about being fanciful, but I couldn't escape the conviction that it knew that it knew things. And to do that it had to know itself.

That crow and I shared an ancestor and a great deal of DNA. My branch of the family had split from the crow's about two hundred million years ago, and since then we had taken rather different lines across country. Its brain was built very differently from mine, and yet impeccably refereed papers were increasingly talking about demonstrations of consciousness in crows.[1] Although some biologists dismiss the idea of any consciousness in any animal apart from humans and some other pretty advanced mammals, there are insistent, niggling suggestions the other way. I suspect – and I'm not alone – not only that consciousness is a lot more common than we think, but also that it has evolved entirely independently on many different occasions. If a brain as different from mine as a crow's can be in some sense the seat of consciousness, then all bets are off, and we should be looking for consciousness a long way from the higher primate enclosure.

Consciousness looks like another example of evolutionary convergence – the independent generation of similar solutions to biological problems, and a principle that compels us to the conclusion that evolution is much more constrained in what it can do than is often thought.[2] But it is not just another example: it is an example of colossal, unimaginable strangeness. For not only has no one ever been able to come up with a half-plausible reason why consciousness should confer a selective advantage, but also no one has been able to suggest coherently that consciousness is simply neuronal firing – that it is nothing more than brain processes. Indeed, there are very compelling reasons to conclude exactly the opposite.[3] No wonder Newberg was driven back to concluding (surely one of the least surprising neurobiological conclusions ever) that altered states of consciousness weren't generated by a particular spot in the brain. Not only is 'mind' not a parietal lobe thing, or a temporal lobe thing, or an amygdala thing. It's not even a brain thing. In fact the universe might not be big enough to hold it.

So here's the position: Evolution looks like a garden in which consciousness springs up all over the place. To our incurably teleological eyes it looks almost as if there's a gardener who loves the heady scent of self-awareness, and selects for it. And yet, from the materialist perspective, there's absolutely no point in evolution behaving that way, and from that perspective there's absolutely no way in which it could have done. Here's a wild, thrilling surmise. The universe isn't just a stage on which conscious creatures act out their dramas. The appearance of purpose isn't deceptive. The universe is a garden for growing consciousness. The gardener is interested in both quality and quantity. Most of this book has been about altered states of consciousness, but still about consciousness. The lands into which the mystics wander have an important continuity with the lands in which they live normally – the cloister, the family, or the office. The most important continuity is that the 'I' – the self, the consciousness – is the traveller. Yes, there is the feeling that the mystical journey takes one deeper into a Consciousness that is intimately related to one's own consciousness, but that process doesn't cause one to lose one's self any more than going home to one's family – the biological source of the 'I' – causes one to lose one's self. Indeed, the feeling is in many ways like a homecoming. There's a reconciliation of something with something, or someone with someone.

The spiritual experiences we have surveyed are rolling, exhilarating voyages into a bigger Consciousness: a Consciousness that is the source and the goal of our own consciousness, which is why we feel that it is home, but is supremely distinct from us and jealously concerned to ensure that the distinction is recognised and rejoiced in.

There are sometimes horrible things seen along the road: Rick Strassman's man-eating insects and buggering dinosaurs on DMT; Hancock's malevolent aliens on ayahuasca; the demons cackling from Chartres Cathedral; the absence of love in depression – an absence so near and so solid that it is a presence. But now that I look back at the reports of the horrors, they seem to me to be descriptions of *strangers*; of things that didn't belong. They are not the normal innkeepers or travelling companions. They are bandits, without a proper passport.

I said that evolution was like a garden, but it is also like an engine, propelling biology (as we see from the fecund generation of consciousness) in the direction of greater self-awareness. Only a self-aware organism can respond in love. And that seems to be the plan. The universe is incontinently breeding more and more creatures with the capacity for spiritual experience. It's breeding more and more lovers.

Who Am I?
The Terrible Problem of Consciousness

I am sitting in a library. My feet are on the floor. I am looking across the room towards a girl. She sees my eyes follow her. I think that she thinks that I am lusting after her. In fact, I am not, but she has no way of being sure. She has no idea what is going on in my head. I have no idea what is going on in her head. What is happening on the surface of the most distant star is more accessible to me than the contents of her mind.

If I have a religious experience, or any other sort of experience, it will happen to the person who is sitting in the library and looking at the girl.

The way that I talk about all this tells you that I hold a distinct philosophical position. '*I* am sitting'; '*My* feet'; '*I* think'; '*I* have no idea.' I plainly believe that there is some sort of creature inhabiting my body; looking out through my eyes; being carried round on my feet. The creature owns the feet and the eyes. It can order them around. The creature does the thinking. It is distinct from the world and from the girl. It assumes that the body of the girl is similarly possessed and commanded by an entity like itself.

The species of creature that lives in me, using 'my' anatomical hardware to feed it, carry it and collect data from the outside world, has been given lots of different names. They include 'soul', 'mind', 'self', 'consciousness' and 'spirit'. The names are used inconsistently and confusingly, and it is not surprising.[1]

Nobody really has any idea at all about what sort of creature it is. Nobody knows where it lives. Nobody has been able to see it, detect it using the most sensitive detectors we have, or weigh it. And that, as we will see, is not for want of trying. Nobody knows what it is made of, or whether it can be killed.

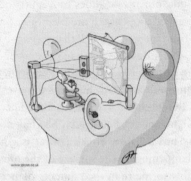

The 'Cartesian theatre' – the idea that there is a real 'I' that sits looking out at the world through its sense receptors.

I am in the philosophy section of the library. The books on the shelves that surround me label my philosophical position 'Cartesian dualism', from the philosopher Descartes who famously and epigrammatically declared, 'I think, therefore I am.' Few declarations beg more questions, and more fundamental questions. The overwhelming consensus in the modern philosophy books is that this Cartesian dualism is wrong. Writers like Daniel Dennett snigger at the naivety of my position, talking dismissively about the 'Cartesian theatre' – the idea that there is a real 'I' that sits in the stalls eating its metaphysical popcorn, looking out at the world's performance through its sense receptors. And yet Dennett's books are liberally sprinkled with personal pronouns. I'm sure that he talks about 'his' wife and 'his' bowels. We all talk in an identically Cartesian way. And we don't just talk about 'I' in the present moment; we talk about it in the past and the future as well. We assume that the 'I' that was there last week and last year is the same one that exists now and will, all being well, be there tomorrow. My body took my 'I' to London yesterday. Our bodies change; our hair drops out; our personalities transmute; our preferences evolve; we have experiences which are filed in our memories and sometimes lost from there. But the 'I' that owns the hair, the preferences and the memories doesn't change. It squats there, somewhere inside us, wholly sovereign, apparently immutable, but incomparably elusive. Only in some very strange and very rare states, induced by extraordinary activity, inactivity or powerful drugs, does the 'I' seem to be nudged off its throne.

That dethronement is sometimes said to be the goal of all religious activity. It is accompanied by ecstasy (although whose ecstasy?), tears (from whose eyes?) and the end of all longing (but who says so?).

During working hours, most mainstream philosophers denounce dualism. But the nine-to-five mockers become dualists as soon as they step outside the philosophy department. They can't help it. If the 'I' is an illusion, generated by neurological smoke and mirrors, it has taken us all in. It rules the way we talk, think, plan, reminisce, love, hate and write books about its illusory nature. It governs every nuance of our lives, and the way we think about our deaths. It survives torture, rape and totalitarianism. It might even survive death. Lots of people certainly behave as if it does. Not only does it seem to be hardwired into us, it seems to be the only thing that is hard-wired into absolutely everyone. Everyone behaves as if the 'I' and (for lovers) the 'you' are the most important and most incontestably real things in the world.

So why all the fuss? Why all the high-brow scepticism about the bloomin' obvious?

The flippant answer is that you don't get a PhD for stating what everybody, everywhere, knows is obviously true. Nor have you ever, since the mind-body problem was first brilliantly articulated by some very clever Greeks a very long time ago. 'The theories that are not counter-intuitive are simply wrong,' says Dennett. To which the cynic might add, 'The theories that are intuitive don't get you tenure.' In the world of professional philosophy, if you're a Cartesian dualist you might as well be the plumber. It is a world ruled by distinctly unclad emperors.

But it is not at all as simple as that. The difficulties are real, salutary and instructive. They teach us that our intuitions have to be listened to very critically. Intuitions shouldn't start off in pole position, ahead of all the other evidence.

If a patient has a massive stroke, causing catastrophic damage to the right hemisphere, the left side[2] of the world can, for most practical purposes, cease to exist for them. The patient might apply eye-shadow only to her right eyelid, or ignore her son unless he stands on her left side. Where now is the over-arching 'I'? Why is 'it' not saying, 'This is ridiculous: my left eyelid is bare'? Or, 'I live in a world which has a left and a right side. Not as much information as before is reaching me from the cameras that I call my eyes, but I can integrate what there is in order to produce a picture of the world that corresponds to the one in which my memories tell me I live.' We don't know why this doesn't happen, but it doesn't.

One of the reasons why we are convinced of the reality of the 'I' is that 'it' seems to have some continuity. The person who experienced London yesterday seems to be the same person who is experiencing Oxford today. But what if the person in Oxford has no recollection of being in London? What if (as some urge is desirable for all of us) he lives so thoroughly in the present that he has no idea that anything other than the present 'I' has ever existed? Disease can maroon people on certain islands of time, preventing the creation of any new long-term memories and leaving the patient with the long-term memories that were entrenched at the time the disease struck.

Jimmie G., aged forty-nine, an ex-submariner in the US Navy, was admitted to a home for the aged near New York City in 1975. The referral letter said, 'Helpless, demented, confused and disorientated.' He came into the consulting room of the neurologist Oliver Sacks. 'Hiya, Doc!' he said as he came in. 'Nice morning! Do I take this chair here?' He was friendly and forthcoming. He answered in detail questions about his childhood and adolescence, and about how, in 1943, he was drafted. His memory of the war was colourful and exciting. But in 1945 his recollection stopped. Sacks noted that when Jimmie moved from his school days to his navy days, the tense shifted from the past tense to the present.

'What year is it now?' asked Sacks.

'Forty-five, man,' came the slightly aggrieved reply. 'What do you mean? We've won the war, FDR's dead, Truman's at the helm. There are great times ahead.'

Sacks asked Jimmie how old he was. Jimmie seemed uncertain. He was calculating. 'I guess I'm nineteen, Doc,' he said at last.

Sacks then did something for which he has never forgiven himself. He pushed a mirror towards Jimmie. 'Tell me what you see,' he said. 'Is that a nineteen-year-old?'

Jimmie looked, turned white, and panicked. 'Christ, what's going on? What's happened to me? Is this a nightmare? Am I crazy?'

Sacks tried to reassure him, and took him to the window. Outside there were children playing baseball. Jimmie watched, and began to collect himself. Sacks left, but came back two minutes later. Jimmie swung round to greet him. 'Hiya, Doc!

Nice morning! You want to talk to me – do I take this chair here?' He had no memory of meeting Sacks just moments before.

'He is, as it were, isolated in a single moment of being,' Sacks wrote in Jimmie's medical notes, 'with a moat or lacuna of forgetting all round him ... He is [a] man without a past (or future), stuck in a constantly changing, meaningless moment ... The remainder of the neurological examination is entirely normal. Impression: probably Korsakov's syndrome, due to alcoholic degeneration of the mamillary bodies.'[3]

Where and what was Jimmie's 'I'? He had no yesterday. If there was an 'I' yesterday, had it not vanished, to be created again and again in the rolling moments? What or who was Jimmie?

Jimmie's was an extreme case. Perhaps one should not generalise too wildly from pathology. But is it really so different for any of us? Whenever we are asked the question, 'Are you conscious?' the answer, if we are awake and neurologically normal, will be an outraged or amused, 'Of course!' But are you conscious when the question is not being asked? The psychologist Susan Blackmore has meditated long and hard on the question, 'What was I conscious of a moment ago?' So have I. It is an illuminating question.

Imagine yourself sitting cross-legged in a Zen meditation hall. It is slightly hot and you are struggling to keep your mind clear of distractions. You don't go to sleep, but suddenly the question, 'Am I conscious?' is dragged back to the forefront of your mind. The 'I' sits up with a jolt, and says that it is. Then comes the rejoinder: 'What was I conscious of a moment ago?'

Perhaps the honest answer is that you are not conscious of having been conscious then at all, but that some memories might possibly be retrieved if you are interrogated hard enough. Retrieval is much more likely if a question suggests the answer – for instance if you are asked, 'Did the student on the far side of the hall sneeze?' We notice the same thing all the time in real life. Suppose a clock is striking ten. You first 'consciously' note it striking when it is on its third strike, but you may very well, on asking the question, 'Is that the first strike?' be able to say, 'No, actually: although I didn't note it at the time, there were two previous sounds.'[4] Or take a police crime re-enactment. Suppose someone has been stabbed to death on the high street by someone dressed in a distinctive striped coat. No one saw the crime itself, and no one remembers seeing the suspect at all. An actor, dressed in a similar coat, walks along the route that the suspect must have taken. People who were originally certain that they remembered nothing manage, after seeing the actor, to drag pictures of the real suspect out from the dusty pigeonholes of their minds.

What is happening? We intuitively want to hang on to the traditional idea of the stream of consciousness. Experiences like this, say Blackmore and many others, deal a death blow to this idea. It is tempting to think that there is a continuous 'me' that is there all the time, perhaps running in the background like a computer program whose icon is not visible or not noted, but that notion has to go. The brain, ex post facto, when it is asked a question about what it was doing in the periods between observed episodes of consciousness, concocts memories to connect the episodes of apparent consciousness. Isolated experiences, fed to our neurones by our sense receptors, get strung together by the clever, story-telling brain to generate an illusion of a coherent, consistent tale. The tale (or its narrator, or principal character) is what we speak of as our self or our consciousness. It's an ingenious novel, but we shouldn't believe in the truth of the tale or the existence of the characters any more than we should believe that Rupert Bear is history.

What are you, on this view? 'You', in the sense that you tend to think of yourself, don't exist at all. 'You' are a string of

beads. The beads are your experiences. The string is composed
of memory and spun on the same wheel that spun the *Iliad* and
the idea of personal immortality. But the string of beads, and
accordingly 'you', are nothing but a metaphor. There is a body,
to which things happen. Even the body, or most of it, is not
composed of the same cells that you were born with. You are
not even biologically continuous with what you describe as your
body, save to the limited extent that your DNA, barring muta-
tions, continues to code for basically the same proteins as it did
when you were at your mother's breast. And even then, bits of
your DNA are switched on and off by processes that are still
wholly mysterious. The things that happen to the body, and the
other data processed by your neurological hardware and software, are stored as 'experiences'. There is a burning imperative, rooted deep in our software, which insists that experiences
necessarily imply an experiencer, and (partly perhaps because
similar hardware generates similar experiences) accordingly
insists on the myth of a continuous consciousness.[5]

The brain, says Dennett, produces from all the information
beamed into it many 'drafts' of what it perceives as 'reality'.[6]
When it is asked a question it rummages around and produces
the draft which seems to provide the best answer. If there were
a real, conscious 'I' in the witnesses to the murder, the 'I', on
first being asked the question, would state that of which it was
conscious. The fact that there was no consciousness which
initially encompassed the man in the distinctive coat means that
there is no consciousness at all.

All our intuition screams at this. To which Dennett and
Blackmore calmly say, 'Well, it would, wouldn't it? The fact that
it screams is precisely the *problem* of consciousness that we're
discussing.'

The traditionalists' typical rejoinder is 'free will'. It is taken
as a given that Daniel Dennett is free to choose whether or not

to write books about the mind-body problem. Susan Blackmore is free to choose whether or not to stroke her cats. I am free to leave the library and go to drink coffee in Blackwells. If it is right to say that I am free, 'I' must mean something. It must mean that there is some Cartesian director that tells my fingers to stop typing, my legs to haul me upright, and my navigational apparatus to steer me round the corner. All our presumptions about ourselves, the world and morality are based on that view. Even the most tightly caged prisoner is free in most respects. He can choose to scratch his nose, hold his breath or think murderous thoughts about his captors.

Or can he? We may be enslaved to something or someone who is always significantly ahead of us.

The Californian physiologist Benjamin Libet did a series of experiments on human patients whose brains had been exposed for surgery. Patients indicated when they decided to move their wrist. There is a fair amount of wiring between the brain and the wrist, and, unsurprisingly, the wrist moved about 200 milliseconds after this decision was apparently made. But here's the really, really odd thing. The brain processes involved in planning the movement could be detected using electrodes. And they seemed clearly to have begun more than a third of a second before the patient became aware of the decision to move. In other words, the decision to move seemed not to have been made by the conscious 'I' at all.[7] More recent work has suggested that this period might in fact be as long as ten seconds.[8]

Just pause to shudder at the consequences of this. If the findings really mean what they appear to say, the universe is a dramatically different place from the one it seems to be.

The enslavement was not complete. The patients apparently retained a right of veto over an action that had been decreed by the mysterious controller. 'If one doesn't have free will, one does have free won't,' says Susan Blackmore.[9]

These objections to the traditional Cartesian dualism come from people who have understood well the nature of the problem of consciousness. But most objections don't. The most widely broadcast objection is the simple materialist sneer – a sneer that asserts that consciousness *is* the physical processes that go on inside the brains of apparently conscious beings. Here is Francis Crick:

> 'You', your joys and your sorrows, your memories and your ambitions, your sense of personal identity and free will, are in fact no more than the behavior of a vast assembly of nerve cells and their associated molecules. As Lewis Carroll's Alice might have phrased: 'You're nothing but a pack of neurons.'[10]

But sneers and assertions do not amount to argument, and those particular sneers and assertions merely beg the very question they purport to answer, namely, 'What is consciousness?'

What are we to make of it all? Is there any answer to the compelling evidence that Charles Foster is a self-generated myth; an onanistic metaphor; a string of inaccurate recollections linked by a flimsy fiction of self? Can I say that anything at all is me? Must I surrender the most powerful convictions that I have – that I exist and can choose to scratch my nose – to the findings of the physiologists?

Susan Blackmore is in no doubt:

> It seems we have some tough choices in thinking about our own precious self. We can hang on to the way it feels and assume that a persisting self or soul exists, even though it cannot be found and leads to deep philosophical troubles. We can equate it with some kind of brain process and shelve the problem of why this brain process should have conscious experiences at all, or we can reject any persisting entity that corresponds to our feeling of being a self.

I think that intellectually we have to take this last path. The trouble is that it is very hard to accept in one's own personal life. It means accepting that there is no one who is having these experiences. It means accepting that every time I seem to exist, this is just a temporary fiction and not the same 'me' who seemed to exist a moment before, or last week, or last year. This is tough, but I think it gets easier with practice.[11]

I respectfully disagree. There are six main reasons.

First, the primary objection to the idea of a self/soul/occupant of the Cartesian theatre is simply that you can't see it, weigh it or otherwise detect its presence, or propose a mechanism for how, if it exists, it interacts with the brain or the world.

This is a rather quaint objection. In the early days of serious scientific investigation of the paranormal, investigators weighed bodies shortly before and immediately after death. The idea was that if the soul existed, it must have a detectable weight, and accordingly the dead body would be lighter than the living. Using appalling experimental methods, the soul was confidently declared to weigh about an ounce. Later workers used more sophisticated methods, including detectors of magnetic fields, temperature, UV and infrared light, but detected nothing.[12] What they really needed to improve was the sophistication of their assumptions, not the sophistication of their apparatus. Do you really expect to see the soul on a PET scan? Do you really imagine consciousness as a substance secreted like acetylcholine at neuronal junctions? Thinking that you can weigh or see everything that exists isn't science, but scientism.

Second, many of the objections conflate memory and consciousness. They are not the same, nor are they mutually dependent. To say that I can't immediately retrieve a document is not the same as saying that I was unconscious at the time it was delivered. The witnesses who saw the man in the distinctive coat weren't unconscious when they were walking down the high street. To be 'conscious of X' means something very different from being simply 'conscious'. The fact that I can't give my own interrogating mind an emphatic answer to the question, 'What were you conscious of a moment ago?' doesn't indicate anything other than that the subject of my consciousness a moment ago hasn't for some reason been stored in such a way as to be immediately accessible. Jimmie G. was no less Jimmie G. by being deprived of some of the hinterland of experience in which most of us dwell for much of the time. The language of that metaphor exposes Blackmore's fiction. Our fundamental experiences are buildings that we often inhabit. Our less fundamental experiences are the pictures on the wall or the trinkets on the shelves. But *we* live in the houses. *We* gaze at the pictures and tell *ourselves* stories. *We* dust our trinkets. But we don't think that the walls, paintings or knick-knacks are *us*. Jimmie G.'s forceful, particular personality and his inalienable conviction that he was Jimmie G. tell us that we are right. Yes, he thought he was nineteen. In fact, he was forty-nine. So what? Those are just numbers. He was never in any doubt that he was himself. And if he didn't doubt, why should we think that memories – the stories we tell ourselves about ourselves – are defining components of what we are?

My three-and-a-half-year-old son Tom likes nothing better than to be told stories about himself. 'Please, Daddy, you tell me a story about Tom.' He will often specify what he would like Tom to do in the stories. Often the commission involves desperately dangerous foreign travel, vast mountains, castles bristling with battlements, crawling with snakes and guarded by dragons, and moats full of sharks. He can repeat the bedtime story almost word-perfectly at breakfast. In one sense he lives the stories as they are told. The acts of derring-do are fascinating to him because, and only because, they were 'done' by him. And yet he is in no danger of thinking that he really did them. He doesn't define himself by reference to them. If he had no more stories about Tom, or had never had any stories about Tom, he would be no less him.

It is the same with the memories of things that really happen. We can get by without our memories, with undiminished selves, just as we can get by without our hair or our arms. Perhaps, if anything, our memories dilute us. Surely a person who is *just* memory is less himself than a person who, like Jimmie G., was pathologically forced to exist in the moment. Children have fewer memories than adults, and yet children are more intensely alive – are more intensely themselves – than those who, when asked to say what makes them tick, simply recite their life story.

So we can respond to Dennett's multiple drafts theory: it might be true. Chiming clocks and murder suspects certainly suggest it. But so what? It is hardly an attack on the notion of consciousness itself. It relates only to the *contents* of consciousness – what we notice about what our conscious brains receive. If anything, by indicating that consciousness is a multi-laminated thing, it deepens the mystery, making the *fact* of consciousness or apparent consciousness more impressive.

Third, the loss by the stroke patient of half of the world isn't loss of half of himself. If anything, 'hemifield neglect' makes the very point it is wheeled out to disprove. It shows that you can lose a lot of things that you might think are fundamental to self-definition without having the slightest effect on it. What has been lost is an awareness of a significant part of the patient's body, and an apprehension of part of the world outside that body. But the 'I' remains unaltered. That patient's conversation will still be full of personal pronouns; he will still have entirely subjective experiences. Going back to the Cartesian theatre: a patient with hemifield neglect is like a spectator in a theatre whose west wing has fallen in. But the spectator is uninjured and unchanged. The show goes on.

Fourth, Libet's delay and the more recent work of Soon and others. Both the experimental method and the interpretation of the results are violently controversial. The details of the method- ological objections are technical and outside the scope of this book, but you don't have to be a neurophysiologist to ask intel- ligently, 'Can you really *time* a conscious experience?' Consciousness of something isn't an all-or-nothing thing, either there or not. Libet himself took his results as evidence for the proposition that whatever mind/self are, they are not the same as the processes in the brain with which they are more or less correlated. And that is surely the natural reading. If you take Libet seriously, you might be led into a form of mystical Buddhism, believing that everything is a consequence of every- thing else, but you can be no prosaic materialist.

Fifth, why did consciousness arise? If you're a straightfor- ward materialist, of course, Darwin has to supply the answer. An answer from any other quarter will get you tarred and feath- ered. But Darwin really can't help. There have been strenuous attempts to recruit him by saying, in the amorphous way derided by Darwinists in anyone else, that consciousness is all part of the mix (which includes theory of mind, language, intelligence) that enables sociability and therefore evolutionarily advanta- geous co-operation. But in talking about consciousness we are talking about purely subjective experiences. And such experi- ences are surely never visible to the rather crude eyes of natural selection. If Libet is right, of course, these experiences come too late to affect any action at all. One might argue that sensi- tive, introspective people could use the insights they gain from looking at themselves to understand others better – and so to manipulate, toady, submit or do whatever will maximise the chances of their own genes being represented in the next generation. And indeed, one can see how, in a society that was already very complex, that might confer an advantage. But it is impossible to conceive how it was ever selected for in the first place. If it confers an advantage sufficiently obvious to be selected for, one might expect it to have emerged more often than it has. Consciousness has only been demonstrated conclusively in *Homo sapiens*,[13] and yet co-operation and community are ubiquitous and very, very ancient.[14]

And sixth, our subjective experiences are, as far as subjective things can be, corroborated by the subjective experiences of others, as related to us.

Back in the library, someone has just scraped a chair on the floor. I heard it. Or I say that I did. The 'I' bit of me, relying on the inputs from its sense receptors, concludes that a chair scraped. Is it seriously to be doubted that there is some objective sense in which the chair scraped and caused a compression of the air between it and me which set off electronic impulses along my auditory nerves? No. Is my attribution of the sound to a scraping chair (an attribution no doubt shared by everyone else in the room, as I would discover were I rude enough to interrupt their work and question them) really suspect? Hardly. The correlation of the sound with the movement of the chair on the lino is eminently reproducible. Yes, the sound will no doubt mean slightly different things to each one of us here. It will evoke different resonances in our heads, depending on our experiences of scraping chairs. So yes, there's a genuinely subjective element. But is it really of a wholly different quality from the agreed diagnosis of 'scraping chair'? The objective bits overlap and correlate perfectly in time, nature and everything else with the subjective bits. The many different 'I's in the room all appear to have had subjective experiences substantially similar, as far as we can tell, to mine. The same is found whenever human beings compare notes about anything. The fact that not every nuance of subjective experience is accessible doesn't mean that we can conclude nothing at all from the plain fact of shared experience. The fact that the girl I am not lusting after would agree with me that the chair has scraped goes some way towards reassuring me that I can talk meaningfully about 'I', and that she can sensibly talk about her.

There are, then, good reasons to be unfashionably, unacademically confident that what every shred of human intuition shrieks to us is right. We can ignore the scoffing of the sceptics who disbelieve in their own existence, and continue to enjoy the splendid show in the Cartesian theatre.

Select Bibliography

Atran, S., *In Gods We Trust: The Evolutionary Landscape of Religion* (New York: Oxford University Press, 2002).

Austin, J., *Zen and the Brain: Towards an Understanding of Meditation and Consciousness* (Cambridge, MA/London: MIT Press, 1998).

Austin, J., *Zen-Brain Reflections: Reviewing Recent Developments in Meditation and States of Consciousness* (Cambridge, MA/London: MIT Press, 2006).

Baker, J. R., 'Psychedelic Sacraments', *Journal of Psychoactive Drugs*, 37(2), 2005, pp. 179–87.

Barrett, J. L., *Why Would Anyone Believe in God?* (Lanham, MD: AltaMira, 2004).

Blackmore, S., *Beyond the Body: An Investigation of Out-of-the-Body Experiences* (London: Heinemann, 1982).

Blackmore, S., *Consciousness: A Very Short Introduction* (Oxford: Oxford University Press, 2006).

Boyer, P., *Religion Explained* (New York: Basic Books, 2002).

Cohen, S., *The Beyond Within: The LSD Story* (New York: Athenaeum, 1964).

Conway Morris, S., *Life's Solution: Inevitable Humans in a Lonely Universe* (Cambridge: Cambridge University Press, 2003).

Crick, F., *The Astonishing Hypothesis: The Scientific Search for the Soul* (New York: Charles Scribner's Sons, 1994).

Dawkins, R., *The God Delusion* (London: Bantam, 2006).

De Waal, F., *Primates and Philosophers: How Morality Evolved* (Princeton: Princeton University Press, 2009).

Dennett, D. C., *Consciousness Explained* (London: Penguin, 1991).

Dennett, D. C., *Breaking the Spell: Religion as a Natural Phenomenon* (London: Penguin, 2007).

Devereux, P., *The Long Trip: A Prehistory of Psychedelia* (London: Penguin, 1997).

Eliade, M., *Shamanism: Archaic Techniques of Ecstasy* (New York: Routledge and Kegan Paul, 1972).

Forman, R. K. C. and J. Andresen (eds), *Cognitive Models and Spiritual Maps: Interdisciplinary Explorations of Religious Experience* (New York: Imprint Academic, 2000).

Forte, R., *Entheogens and the Future of Religion* (San Francisco: Council on Spiritual Practices, 1997).

Fox, M., *Religion, Spirituality and the Near Death Experience* (London: Routledge, 2003).

Halifax, J., *Shaman: The Wounded Healer* (New York: Crossroad, 1982).

Hancock, G., Supernatural: Meetings with the Ancient Teachers of Mankind (London: Arrow, 2006)

Harner, M., *The Way of the Shaman* (New York: Bantam, 1982).

Horgan, J., *Rational Mysticism* (Boston/New York: Houghton Mifflin, 2003).

Huxley, A., *The Doors of Perception and Heaven and Hell* (Harmondsworth: Penguin, 1959).

James, W., *The Varieties of Religious Experience* (New York: Macmillan, 1961).

Jansen, K. L. R., 'Ketamine, Near-Birth and Near-Death Experiences', in *Ketamine: Dreams and Realities* (Florida: Multidisciplinary Association for Psychedelic Studies, 2001), pp. 92–166.

Joyce, R., *The Evolution of Morality* (Cambridge, MA: MIT Press, 2007).

Letcher, A., *Shroom. A Cultural History of the Magic Mushroom* (London: Faber and Faber, 2006).

Lewis-Williams, J. D., *The Mind in the Cave: Consciousness and the Origins of Art* (London: Thames and Hudson, 2002).

Lewis-Williams, J. D. and D. Pearce, *Inside the Neolithic Mind: Consciousness, Cosmos and the Realm of the Gods* (London: Thames and Hudson, 2005).

McGilchrist, I., *The Master and His Emissary: The Divided Brain and the Making of the Western World* (London: Yale University Press, 2009).

McKenna, T., *The Archaic Revival* (San Francisco: HarperSanFrancisco, 1991).

McKenna, T., *Food of the Gods: The Search for the Original Tree of Knowledge. A Radical History of Plants, Drugs and Human Evolution* (New York: Bantam, 1993).

McKenna, T., *True Hallucinations: Being an Account of the Author's Extraordinary Adventures in the Devil's Paradise* (San Francisco: HarperSanFrancisco, 1993).

Mithen, S., *The Prehistory of the Mind: A Search for the Origins of Art, Religion and Science* (London/New York: Thames and Hudson, 1996).

Newberg, A. and D'Aquili, *Why God Won't Go Away* (New York: Ballantine, 2001).

Newberg, A. and M. R. Waldman, *Born to Believe* (New York: Free Press, 2006).

Ott, J., *Pharmacotheon: Entheogenic Drugs, Their Plant Sources and History* (Kennewick, WA: Natural Products Co., 1996).

Ott, J., 'Entheogens II: On Entheology and Entheobotany', *Journal of Psychoactive Drugs*, 28(2), 1996, pp. 205–9.

Partridge, C., 'Sacred Chemicals: Psychedelic Drugs and Mystical Experience', in C. Partridge and T. Gabriel (eds), *Mysticisms East and West: Studies in Mystical Experience* (Carlisle: Paternoster Press, 2003), ch. 7.

Pearson, J. L., *Shamanism and the Ancient Mind: A Cognitive Approach to Archaeology* (Walnut Creek: AltaMira, 2002).

Pinchbeck, D., *Breaking Open the Head: A Psychedelic Journey into the Heart of Contemporary Shamanism* (New York: Broadway Books, 2002).

Ramachandran, V. S. and S. Blakeslee, *Phantoms in the Brain: Probing the Mysteries of the Human Mind* (London: Fourth Estate, 1998).

Ruck, C. A. P., J. Bigwood, D. Staples, J. Ott and R. Gordon Wasson, 'Entheogens', *Journal of Psychedelic Drugs*, 1(1–2), 1979, pp. 145–6.

Sabom, M., *Light and Death: One Doctor's Fascinating Account of Near-Death Experiences* (Grand Rapids, MI: Zondervan, 1998).

Saunders, N., A. Sanders and M. Pauli, *In Search of the Ultimate High. Spiritual Experience through Psychoactives* (London: Rider, 2000).

Schultes, R. E. and A. Hofmann, *Plants of the Gods: Their Sacred, Healing and Hallucinogenic Powers* (Rochester, VT: Healing Arts Press, 1992).

Shanon, B., *The Antipodes of the Mind: Charting the Phenomenology of the Ayahuasca Experience* (Oxford: Oxford University Press, 2002).

Sloan Wilson, D., *Darwin's Cathedral: Evolution, Religion and the Nature of Society* (Chicago: University of Chicago Press, 2003).

Smith, H., *Cleansing the Doors of Perception: The Religious Significance of Entheogenic Plants and Chemicals* (Boulder, Colorado: Sentient Publications, LLC, 2003).

Stace, W. T., *Mysticism and Philosophy* (New York: Tarcher, 1987).

Suber, E. and D. Sloan Wilson (eds), *Unto Others: The Evolution and Psychology of Unselfish Behaviour* (Cambridge, MA: Harvard University Press, 1999).

Tart, C. (ed.), *Altered States of Consciousness* (New York: HarperCollins, 1990).

Taves, A., *Fits, Trances and Visions: Experiencing Religion and Explaining Experience from Wesley to James* (Princeton: Princeton University Press, 1999).

Wallis, R. J., *Shamans/Neo-Shamans: Ecstasy, Alternative Archaeologies and Contemporary Pagans* (London: Routledge, 2003).

Wasson, R. Gordon, *The Wondrous Mushroom: Mycolatry in Mesoamerica* (New York: McGraw-Hill Book Company, 1980).

Wasson, R. Gordon, S. Kramrisch, J. Ott and C. Ruck (eds), *Persephone's Quest: Entheogens and the Origins of Religion* (New Haven/London: Yale University Press, 1986).

Weil, A., 'Review of *Persephone's Quest: Entheogens and the Origins of Religion*', *Journal of Psychoactive Drugs*, 20, 1988, pp. 489–90.

Notes

Chapter 1: Matter Matters:
Religious People are Made of Molecules too

1 So a word to them. The ancient Christian position – one for which the early church fought tooth and nail against the Gnostics – is that there is an essential solidarity between, if not an identity of, matter and spirit. The Gnostics said that what mattered was pure spirit. Material things were nasty and dirty and got in the way of the business of the spirit. The incarnation was a profound embarrassment to them, and so they rejected it. Jesus wasn't *really* human: he just looked human. If he appeared to eat, get tired and defaecate, the appearance was an illusion. He didn't really die: spirit can't be killed. In *The Da Vinci Code* (London: Bantam, 2003), Dan Brown, of course, got things completely the wrong way round. It was the Gnostics, not the orthodox Christians, who would have had a real problem about Jesus being married. Sex was really filthy, earthy and unspiritual. The Christians had no such hang-ups. A married Jesus would have presented no problems at all for wine-supping, steak-chewing, merrily procreating, loudly carousing, properly incarnational Christians. It's just that, unusually for a Jewish man of his age and time, Jesus doesn't seem to have been married. He heartily endorsed life in all its spiritual–material fullness. The incarnation is God's resounding 'yes' to matter. The notion that 'John Brown's body lies a-mouldering in the grave, but his soul goes marching on' is pure Gnosticism with a Greek flavour. In Christianity, John Brown's

mind-body-soul unity gets raised from the grave, walks, runs, eats and drinks, and the heavier-and-more-solid-than-before resurrection body of Jesus is the first indication of what it will look like.

All of which is a roundabout theological way of saying that, for Christianity at least, matter matters, profoundly and for ever. The material world isn't just going to be tossed into the flames, but restored. If matter matters in this sense, one might expect interference with matter to have theological consequences now. And indeed it seems that it does.

The same point is made superbly by John Horgan, without a single Christian allusion. He is expressing his misgivings with the idea of *advaita* – mystical oneness with the Universe – an idea to which we return repeatedly in this book: 'I admit that, as a result of my entheogenic experience in 1981, I still find oneness metaphysically creepy. I keep thinking about the nirvana paradox: If nirvana is so great, why does God create? James Austin's first koan asked: "When all things return to the one, where is the one returned to?" Good question. The reduction of all things to one thing is arguably a route to oblivion; one thing equals nothing. I am hardly alone in fretting over this dilemma. The Hindu sage Ramakrishna no doubt had it in mind when he said, "I want to taste sugar; I don't want to be sugar"' (*Rational Mysticism*, Boston/New York: Houghton Mifflin, 2003, p. 230).

2 See Chapters 2, 4 and 9.

3 A thesis of James Austin in *Zen and the Brain: Toward an Understanding of Meditation and Consciousness* (Cambridge, Mass./London: MIT Press, 1998) and *Zen-Brain Reflections: Reviewing Recent Developments in Meditation and States of Consciousness* (Cambridge, Mass./London: MIT Press, 2006). The thesis is well summarised by John Horgan in *Rational Mysticism*, pp. 125–6: 'Apraxia [for instance] undercuts the ability to initiate complex actions. In contrast, mystical awakening gives you a "promethean hyperpraxia", an ability to translate intention into action more rapidly, effortlessly, and creatively. People who suffer from prosopanosia cannot consciously recognise faces, even though

lie detectors indicate that they are responding to the faces on an unconscious "gut" level. In kensho, Austin commented, the reverse may be true; that is, you retain your ability to recognise what you are seeing, but you no longer respond to it on a personal, emotional level. Then there is a bizarre disorder called multagnosia, in which you can perceive objects in a scene only individually, not as parts of a whole. Zen, in contrast, boosts the ability to see holistically.'

4 A. Newberg and D'Aquili, *Why God Won't Go Away* (New York: Ballantine, 2001), p. 89.

5 The extraordinary work of Libet and his successors, which is highly pertinent to the question of whether we have free will, is dealt with in detail in the Appendix (pp. 261–7).

6 See A. Newberg and M. R. Waldman, *Born to Believe* (New York: Free Press, 2006), pp. 161–2.

7 Not always, of course. It depends what damage the tumour has done. See ibid., p. 162.

8 Or, if you prefer, men think less fluently than women about thoughts, feelings, and their own and others' minds.

9 'Theory of Mind' is broadly the ability to think oneself into the mind of another; to put oneself in another's shoes.

10 The merciful paralysis is courtesy of the pons, a structure on the brain stem.

Chapter 2: God Head: The Anatomy of Religion

1 'Peet', *A brain structure dedicated to religious experience*, http://www.paranormal.org.uk (1998), cited in Victoria Powell, *Neurotheology: With God in Mind,* http://www.clinicallypsy-ched.com/neurotheologywithgodinmind.htm (2004).

2 *Heartbeat in the Brain* (1970). Fielding is apparently still a true believer in the practice, and runs the Trepanation Trust: http://www.trepanation.org.

3 The centrality of the pineal in esoteric religion has been recently underscored by Jonathan Black, *The Secret History of the World* (London: Quercus, 2008), pp. 78–80.

4 A. Newberg and M. R. Waldman, *Born to Believe* (New York: Free Press, 2006), p. 178.

5 Examples include aspartate and L-glutamate.

6 J. Austin, *Zen and the Brain: Toward an Understanding of Meditation and Consciousness* (Cambridge, Mass./London: MIT Press, 1998).

7 Austin, ibid.

8 See M. K. Sim and W. F. Tsoi, 'The effects of centrally acting drugs on the EEG correlates of meditation', *Biofeedback and Self-Regulation*, 17:3, 1992, pp. 215–20, cited in J. Horgan, *Rational Mysticism* (Boston/New York: Houghton Mifflin, 2003), p. 129.

9 Many will be (rightly) upset by this simplification. There are many problems with it, but none of the problems makes the main argument void.

10 For further discussion of the relative role of the hemispheres, see the story of Dom Bede Griffiths in Chapter 4.

11 See M. S. Gazzaniga, 'The Split Brain in Man', *Scientific American*, 217, 1967, pp. 24–9. See too M. S. Gazzaniga and R. W. Sperry, 'Language after section of the cerebral commissures', *Brain*, 90, (I), 1967, pp. 131–48; R. W. Sperry and M. S. Gazzaniga, 'Language following disconnection of the hemispheres', in C. H. Millikan and F. L. Darley (eds), *Brain Mechanisms Underlying Speech and Language* (New York: Grune & Stratton, 1967), pp. 177–84; R. W. Sperry, M. S. Gazzaniga and J. E. Bogen, 'Interhemispheric relationships: the neocortical commissures; syndromes of hemisphere disconnection', in P. J. Vinken and G. W. Bruyn (eds), *Handbook of Clinical Neurology* (Amsterdam: North-Holland Publishing Company, 1969), 4, pp. 177–84.

12 The conclusion reached by Sperry. See R. W. Sperry, P. J. Vogel and J. E. Bogen, 'Syndrome of hemisphere deconnection', in P. Bailey and R. E. Foil (eds), *Proceedings 2nd Pan-Am Congress of Neurology* (Puerto Rico, 1970), pp. 195–200. R. W. Sperry, E. Zaidel and D. Zaidel, 'Self recognition and social awareness in the deconnected minor hemisphere', *Neuropsychologia*, 17, 1979, pp. 153–66; R. W. Sperry, 'A modified concept of consciousness',

Psychological Review, 76, 1969, pp. 532–6; R. W. Sperry, 'Mind-Brain Interaction: Mentalism, Yes; Dualism, No', *Neuroscience*, 5, 1980, pp. 195–206; R. W. Sperry, 'Changing Priorities', *Ann. Rev. Neurosci.*, 4, 1981, pp. 1–15.

13 The conclusion of Gazzaniga. See above.

14 As many did: witness the 'Cargo Cults'.

15 Justin Barrett observes that there is a further problem complicating the 'talking hemispheres' explanation for apparent gods: '[I]n most people the left cannot "hear" the right talking (as in self-speak), but can only hear itself because the right doesn't speak. The right can't "hear" the left because the right can't understand language. So the self-speech being mistaken for another consciousness cannot be readily mapped onto the hemispheric hypothesis' (personal communication, 2009).

16 Newberg and D'Aquili put it well: 'What we think of as reality is only a rendition of reality that is created by the brain ... Nothing enters consciousness whole. There is no direct, objective experience of reality. All the things the mind perceives – all thoughts, feelings, hunches, memories, insights, desires and revelations – have been assembled piece by piece by the processing powers of the brain from the swirl of neural blips, sensory perceptions, and scattered cognitions dwelling in its structures and neural pathways' (*Why God Won't Go Away*, New York: Ballantine, 2001, pp. 35–6).

17 *The Doors of Perception and Heaven and Hell* (Harmondsworth: Penguin, 1959).

18 Many neuroscientists would disagree. They would equate processing power with the degree of connectivity – the number of connections – rather than the number of neurones.

19 See Matt. 18:3; 19:14; Mark 10:14; Luke 18:16.

20 The psychologist, Zen Buddhist and former parapsychological researcher Susan Blackmore described to the American writer John Horgan her experience of meditation-induced self-transcendence – a state of permanent mindfulness. She linked it almost casually, but very interestingly, to her relationship to her

children. Horgan writes: 'During one seven-week period when [Blackmore] tried to remain continually mindful, she had difficulty driving. She could not work on page proofs of her writings. Once, walking across a road, she was almost struck by a car. On the other hand her relationship with her two children – then two and four years old – improved. "Kids live in the present moment, and if you live there with them, it's much more alive." When her kids screamed at her, she screamed back. "And then two seconds later they stop screaming and shouting, and then you stop. You don't have any of this grownup stuff: 'I'm still cross.' It's just gone"' (Horgan, *Rational Mysticism*, pp. 116–17).

21 See Newberg and Waldman, *Born to Believe*, pp. 167–90. Newberg's methodology has been questioned. If someone is sufficiently aware to indicate to a researcher when they are at the mountain-top moment, can they really be there? The SPECT scan is only an instantaneous snapshot of the process. There are also some discrepancies between the SPECT results and results obtained using electroencephalography and MRI scanning. In theory, functional MRI scanning should give the best look at the evolving process which a mystical state induced by meditation plainly is, but many practitioners would no doubt feel profoundly inhibited if they had to meditate inside an MRI scanner. Such inhibitions might very well not apply to functional MRI imaging of, for instance, the allegedly mystical states induced by hallucinogenic drugs.

22 This is controversial: see, for instance, the self-explanatorily entitled paper, H. O. Karnath, S. Farber and M. Himmelbach, 'Spatial awareness is a function of the temporal not the posterior parietal lobe', *Nature*, 2001, pp. 950–3.

23 Newberg and Waldman, *Born to Believe*, p. 176.

24 Cf. 1 Cor. 13:12.

25 Newberg and Waldman, *Born to Believe*, pp. 178–9.

26 Ibid., p. 179.

27 He is, however, the best known. That is probably because he has chosen to publish (and has published so successfully) in the popular

scientific press as well as in peer-reviewed journals. His fame should not suggest that his conclusions are necessarily the most scientifically sustainable. A comprehensive audit of the evidence about the neurological basis for religious experience by Jensine Andresen indicated that you can find a scientifically reputable authority for almost every proposition in the field of neuro-theology. Danish researchers, for instance, used SPECT scans to examine brain activity in meditators, and concluded – flatly contradicting Newberg – that they showed increased, not decreased, parietal lobe activity and decreased, not increased, frontal lobe activity. See J. Andresen and R. K. C. Forman (eds), *Cognitive Models and Spiritual Maps: Interdisciplinary Explorations of Religious Experience* (New York: Imprint Academic, 2000).

28 Newberg and D'Aquili, *Why God Won't Go Away*, p. 40.

29 The 'flow' experience.

30 Newberg and D'Aquili, *Why God Won't Go Away*, p. 41.

31 Newberg and Waldman, *Born to Believe*, p. 183.

32 Ibid., pp. 183–4.

33 Ibid., p. 184.

34 See Chapter 4.

35 Newberg and Waldman, *Born to Believe*, p. 179.

36 Ibid., p. 180. He goes on: 'But they wouldn't be able to put it into words, because language is a highly interpretative process.'

37 Chapter 3.

38 Cited in Newberg and Waldman, *Born to Believe*, p. 243.

39 Ibid., p. 226.

40 Ibid., p. 244.

41 Yes, the notion of a 'right-brained'/'left-brained' person is an oversimplification, but it is a good enough approximation to the truth to be useful in the present context.

42 Ibid., p. 214. He adds that there is 'one possible exception' to the 'scientific mystery'. One of the Pentecostal subjects had a resting-state scan. She had the unusual asymmetrical thalamic activity very similar to that found in the resting-state scans of the monks and nuns. Newberg thinks that this suggests either that the

Pentecostal was born with an unusual capacity for mystical experience, or that intensive spiritual practice over the years changed her thalamus to make mystical experiences happen more readily.

43 Many of the papers that make headlines in the lay press really make very unsurprising reading. A good example is U. Schjoedt, H. Stødkilde-Jørgensen, A. W. Geertz and A. Roepstorff, 'Highly religious participants recruit areas of social cognition in personal prayer', *Social Cognitive and Affective Neuroscience*, 2009, 4(2):199–207. The abstract reads, inter alia: 'Distinct from formalized praying and secular controls, improvised praying activated a strong response in the temporopolar region, the medial prefrontal cortex, the temporo-parietal junction and precuneus. This finding supports our hypothesis that religious subjects, who consider their God to be "real" and capable of reciprocating requests, recruit areas of social cognition when they pray. We argue that praying to God is an intersubjective experience comparable to "normal" interpersonal interaction.' In short, if you think that you are talking to God, your brain behaves as if you are in a conversational relationship with him. Of course this tells us nothing at all about the reality or otherwise of the entity with whom the supposed conversation is taking place.

44 1 Cor. 14:5.

45 See L. Carlyle May, 'A survey of glossolalia and related phenomena in non-Christian religions', *American Anthropologist, New Series*, 58:1, 1956, pp. 75–96; and G. J. Jennings, 'An ethnological study of glossolalia', *Journal of the American Science Affiliation*, 20:93, 1968. Jennings notes that *glossolalia* is practised amongst some of the peyote-using tribes of North America, the Haida Indians of the Pacific Northwest, some Sudanese shamans, the Shango cult of the West Coast of Africa, the Trinidadian Shago cult, Haitian Voodoo practitioners, some native South Americans and Australian Aborigines, some Inuit and Siberian tribes, the Dyaks of Borneo, the Zor cult of Ethiopia, the Curanderos of the Andes, and several others.

46 N. P. Spanos, W. P. Cross, M. Lepage and M. Coristine, 'Glossolalia

as learned behaviour: An experimental demonstration', *Journal of Abnormal Psychology*, 95(1), 1986, pp. 21–3, cited in Newberg and Waldman, *Born to Believe*, p. 196.

47 It might be suggested that these studies indicate that *glossolalia* isn't language at all, but meaningless babble. But that won't work. The utterances are audibly sufficiently coherent and syntactical to be quite unmistakably language.

48 It ought to be emphasised, though, that extrapolations from brain imaging studies are dubious. The dangers are superbly summarised in Iain McGilchrist, *The Master and His emissary: The Divided Brain and the Making of the Western World* (London: Yale University Press, 2009). Where, though, *absolutely nothing* happens in precisely the areas that one might expect to be blazing away, it is neurologically permissible at least to raise an eyebrow.

49 Newberg and Waldman, *Born to Believe*, pp. 200–1.

50 Powell, *Neurotheology*.

Chapter 3: The Holy Helix: Genetically Predestined to Believe?

1 D. Hamer, *The God Gene: How Faith is Hardwired into our Genes* (New York/London: Doubleday, 2004).

2 See, for instance, the results of the Minnesota Study of Twins Raised Apart. Some of the more recent conclusions are in L. B. Koenig and T. J. Bouchard Jr, 'Genetic and Environmental Influences on the Traditional Moral Values Triad – Authoritarianism, Conservatism and Religiousness – as Assessed by Quantitative Behavior Genetic Methods', in P. McNamara (ed.), *Where God and Science Meet: How Brain and Evolutionary Studies Alter Our Understanding of Religion. Volume I: The Evolutionary Psychology of Religion: How Evolution Shaped the Religious Brain* (Westport, CT: Praeger, 2006).

3 Hamer, *The God Gene*, pp. 211–12.

4 C. Zimmer, 'Faith-boosting genes: A search for the genetic basis of spirituality', *Scientific American*, October 2004.

5 He is Associate Professor of Biology at the University of Minnesota, Morris.

6 http://pharyngula.org/index/weblog/comments/no_god_and_no_god_gene.

7 Ibid.

8 Ibid.

9 Ibid.

10 Ibid.

Chapter 4: Wholly Mad or Holy Madness?

1 I don't discuss in this chapter the notion of 'crazy-wisdom' – the theme of strange, antisocial or other behaviour in so-called mystics which falls short of psychological or neurological disease. The Pauline idea of 'being fools for Christ' is often cited in that context, but seems to me to have no connection with it at all. A good starting point for anyone interested in 'crazy-wisdom' is John Horgan's book *Rational Mysticism* (Boston/New York: Houghton Mifflin, 2003), pp. 51–4.

2 http://www.rethink.org/about_mental_illness/peoples_experiences/blogs/ebony/delusions_are_funny.html.

3 Although she is often called 'St Christina the Astonishing' she has never been formally beatified.

4 *Thomas of Cantimpré: The Collected Saints' Lives: Abbot John of Cantimpré, Christina the Astonishing, Margaret of Ypres, Lutgard of Aywières*, edited and with an introduction by Barbara Newman; translation by Margot H. King and Barbara Newman (Turnhout: Brepols, 2008).

5 See http://saints.sqpn.com/saintc80.htm.

6 There is a superb account of the history of epilepsy in O. Temkin, *The Falling Sickness: A History of Epilepsy from the Greeks to the Beginning of Modern Neurology* (Baltimore: Johns Hopkins University Press, 1994).

7 E. G. White, *Spiritual Gifts: My Spiritual Experience, Views and Labors* (Battle Creek: James White, 1860), vol. 2, p. 37.

8 Before the battle of Milvian Bridge in AD 312, Constantine looked up at the sun and above it he saw a cross and the Greek words, 'By this, conquer.' He ordered his troops to emblazon their shields with the Christian symbol 'Chi-Ro'. They won the day.

9 See Chapter 6.

10 The link between epilepsy and creativity is discussed by Eve LaPlante in *Seized: Temporal Lobe Epilepsy as a Medical, Historical and Artistic Phenomenon* (Backinprint.com, 2000).

11 See J. L. Saver and J. Rabin, 'The neural substrates of religious experience', *Journal of Neurospsychiatry and Clinical Neurosciences*, 9, 1997, pp. 498–510, cited in A. Newberg and D'Aquili, *Why God Won't Go Away* (New York: Ballantine, 2001), pp. 110–11.

12 V. S. Ramachandran and S. Blakeslee, *Phantoms in the Brain: Probing the Mysteries of the Human Mind* (London: Fourth Estate, 1998), Chapter 9. He acknowledges, though, that these assertions about a link between hyper-religiosity and epilepsy are contentious and largely anecdotal.

13 Persinger has made no secret of his belief that religion is pathological. He has suggested that if meditation has a potent effect on you, you might be suffering from low-grade epilepsy.

14 'Alien abduction', *New Scientist*, 19 November 1994, pp. 29–31.

15 See R. Hercz, 'The God Helmet', *Saturday Night*, October 2002, pp. 40–6.

16 J. Horgan, 'Spirit Tech: How to wire your brain for religious ecstasy', 2007, http://www.slate.com/id/2165004.

17 Ibid.

18 J. Horgan, 'The God Experiments', *Discover*, 20 November 2006.

19 Ibid.

20 http://www.nature.com/news/2004/041206/pf/041206-10_pf.html.

21 J. Hitt, 'This is your brain on God', http://www.wired.com/wired/archive/7.11/persinger_pr.html.

22 Hercz, 'The God Helmet'.

23 Horgan, 'The God Experiments'.

24 See Chapter 2.

25 Or 100 per cent, depending on whether you think that Jesus should be included in the statistics. And even he, if the Christians are right, was truly, properly dead.

26 See Chapter 8.

27 See Chapter 9.

28 The eagle is a common symbol in medieval writings of divine grace.

29 The Migne edition, cols 110–11, cited in C. Singer, 'The Visions of Hildegard of Bingen', in *From Magic to Science: Essays on the Scientific Twilight* (London: Ernest Benn, 1928), reprinted in the *Yale Journal of Biology and Medicine*, 78, 2005, pp. 57–82, at p. 77.

30 The Migne edition, col. 384, cited in Singer, ibid., p. 78.

31 See Chapter 6.

32 In the *Zelus Dei* and *Sedens Lucidus*, cited in Singer, 'The Visions of Hildegard of Bingen', p. 78.

33 See Singer, ibid., p. 78.

34 The Migne edition, col. 384, cited Singer, ibid., p. 78.

35 Singer, ibid., p. 78. See too O. Sacks, *Migraine: Understanding a Common Disorder* (London: Duckworth, 1985).

36 J. Jaynes, *The Origin of Consciousness in the Breakdown of the Bicameral Mind* (Boston: Houghton Mifflin, 1990).

37 I am using the expression 'schizoid' here as imprecise shorthand to express a basic mental dividedness. I acknowledge (courtesy of Iain McGilchrist) that schizoid personality type has no direct relationship with schizophrenia. Patients with a schizoid personality tend to be flat, dull and cold, with no particular propensity to hear voices. I acknowledge too, that Jaynes was not saying that Bronze Age men were schizophrenic, but rather that, in McGilchrist's words, 'in the modern world the phenomena of schizophrenia recapitulate the experiences of Achaean man' (personal communication, 2009).

38 C. Higgins, *It's All Greek To Me* (London: Short Books, 2008), p. 87.

39 For further discussion, see Chapter 11.

40 Depicted, for instance, in the book of 1 Samuel.

41 See Chapter 9.

42 T. Posey, 'Auditory hallucinations of hearing voices in 375 normal subjects', *Imagination, Cognition and Personality*, 3, 1983; J. Hamilton, 'Auditory hallucinations in nonverbal quadriplegics', *Psychiatry*, 48, 1988.

43 Iain McGilchrist points out that 'this area is fraught with difficulty. It is true that some [auditory hallucinations] are from the right hemisphere, but some are from the left. Equally, it is more than likely that schizophrenia involves abnormal lateralisation of the brain, so it may be that some schizophrenic voices are not "speaking" from the "other" hemisphere, as Jaynes's theory supposes – just from their normal language hemisphere, which in their case is right rather than left. . .' (personal communication, 2009).

44 This is broadly the criticism levelled by Ned Block. See N. Block, 'Review of Julian Jaynes' Origins of Consciousness in the Breakdown of the Bicameral Mind', *Cognition and Brain Theory*, 4, 1981, pp. 81–3; cf. M. Kuijsten, *Reflections on the Dawn of Consciousness: Julian Jaynes' Bicameral Mind Theory Revisited* (Julian Jaynes Society, 2007), pp. 303–35.

45 Over thirteen years (1980–93), 1,200 tourists with 'severe, Jerusalem-generated mental problems' have been referred to Kfar Shaul. Of these, 470 were admitted. On average, 100 such tourists are seen at Kfar Shaul each year; 40 are admitted. See Y. Bar-El, R. Durst, G. Katz, J. Zislin, Z. Strauss and H. Knobler, 'Jerusalem Syndrome', *British Journal of Psychiatry*, 176, 2000, pp. 86–90.

46 Ibid., p. 86.

47 Ibid., p. 87.

48 Ibid., p. 88.

49 It has long been so. One of the earliest documented examples is that of Margery Kempe, who made her way to Jerusalem in 1413, clad in white to symbolise her sexual purity. After her fourteen children, she had finally renounced sexual relations with her husband. Her three weeks in the Holy Land were psychiatrically

very interesting. At the sites associated with Jesus' suffering and death she had dramatic visions, but the floodgates burst open at Calvary. She began to sob uncontrollably. For the rest of her life the tears returned at times of devotional intensity. Embarrassed by her eccentricity, her fellow pilgrims abandoned her. She travelled slowly back to England, alone, stopping for a long time in Assisi and Rome. In Rome she experienced an almost literal consummation of her devotion – a mystical marriage to the Godhead.

50 Other places can have similar effects. Bar-El (ibid., p. 89) comments, 'Hysterical or psychotic manifestations related to places – such as those that appear at Mecca, holy places in India, Christian holy places where the Virgin Mary is worshipped, and evangelical rallies – may well resemble our description [of the Jerusalem syndrome].' Freud reported having an experience of derealisation at the Acropolis in Athens; see S. Freud, *A Disturbance of Memory on the Acropolis* (1936), reprinted in the *Standard Edition of the Complete Psychological Works of Sigmund Freud*, trans. and ed. J. Strachey (London: Hogarth Press, 1953–74), vol. 22, p. 239, cited in Bar-El et al., ibid., p. 89. Magherini describes the 'Stendhal syndrome' in which art-loving tourists in Florence, seeing famous paintings, were catapulted into a psychotic episode; see G. Magherini, *Syndrome di Stendhal* (Milan: Fettrinelli, 1992), cited in Bar-El et al., ibid., p. 89.

51 These are the classic stages of 'Type III' or 'pure' Jerusalem syndrome, per Bar-El et al., ibid., pp. 88–9. There is an energetic debate in the literature about the role of Jerusalem itself in the genesis of the 'syndrome', and about the classification of the various stages of the 'syndrome'. Some say that the syndrome is merely evidence that established psychotics use Jerusalem as a stage for the florid manifestation of their disease. Thus Kalian and Witztum contend that 'Jerusalem should not be regarded as a pathogenic factor, since the morbid ideation of the affected travellers started elsewhere. Jerusalem syndrome should be regarded as an aggravation of a chronic mental illness, and not

a transient psychotic episode. The eccentric conduct and bizarre behaviour of these colourful yet mainly psychotic visitors became dramatically overt once they reached the Holy City – a geographical locus containing the axis mundi of their religious belief' (M. Kalian and E. Witztum, 'Comments on Jerusalem Syndrome', *British Journal of Psychiatry*, 176, 2000, p. 492).

52 Bar-El et al., 'Jerusalem Syndrome', p. 89.

53 Ibid.

54 A good example is the cult that grew up in the American Midwest in the 1950s. A woman believed that she was getting messages from outer space. She attracted quite a following. One message said that America would be destroyed on a particular date by a catastrophic flood. The group loudly told America to prepare for doom. Doom didn't arrive. But the group, far from disbanding, convinced of the error of its ways, grew in strength and confidence, convinced instead that America had been saved because of the existence of them, the faithful remnant.

55 See C. Foster, *The Selfless Gene: Living with God and Darwin* (London: Hodder & Stoughton, 2009).

56 It can be argued very compellingly that religion has had a disastrous effect on the health of the world as a whole. The Inquisition didn't do much for the peace of mind or longevity of Spanish Jews. The grotesque parody of Islam which is reactionary Wahabism has the blood of thousands of innocents on its hands – as has the right-wing US evangelical response to it.

57 L. Francis and M. Robbins, 'Personality and glossolalia: a study among male evangelical clergy', *Pastoral Psychology*, 51.5, 2003, cited in A. Newberg and M. R. Waldman, *Born to Believe* (New York: Free Press, 2006), p. 197.

58 A. Greeley, 'Mysticism goes mainstream', *American Health*, 6, 1987, pp. 47–9. Other studies have come to similar conclusions. For further discussion, see Newberg and D'Aquili, *Why God Won't Go Away*, pp. 107–13.

59 See Chapters 10, 11, 12 and 13.

60 Gal. 6:17.

61 From a letter of Padre Pio quoted on http://www.ewtn.com/padrepio/mystic/stigmata.htm.

62 Ibid.

63 The Miracle of Damascus, http://www.reu.org/public/soufan/damas.htm.

64 See Chapter 8.

65 http://www.reu.org/public/soufan/damas.htm.

66 J. Wilbert, *Warao Basketry: Occasional papers of the Museum of Cultural History, University of California at Los Angeles*, No. 3, 1975, pp. 5–6.

67 'The stigmata of Heather Woods', http://www.assap.org/newsite/articles/Stigmata.html.

68 '[I]n an unthinking moment, I ... began to indolently study diseases, generally. I forget which was the first distemper I plunged into – some fearful, devastating scourge, I know – and, before I had glanced half down the list of "premonitory symptoms", it was borne in upon me that I had fairly got it. I sat for a while frozen with horror.' He unwisely continued to read. 'I came to typhoid fever – read the symptoms – discovered that I had typhoid fever, must have had it for months without knowing it – wondered what else I had got; turned up St Vitus's Dance – found, as I expected, that I had that too – began to get interested in my case, and determined to sift it to the bottom, and so started alphabetically – read up ague, and learnt that I was sickening for it, and that the acute stage would commence in about another fortnight. Bright's disease, I was relieved to find, I had only in a modified form, and, so far as that was concerned, I might live for years. Cholera I had, with severe complications; and diphtheria I seemed to have been born with' (J. K. Jerome, *Three Men in a Boat*, 1889).

69 In the charismatic Christian tradition, sympathetic pain is sometimes used to indicate that God wants the cause of the pain to be addressed by prayer. Thus, for instance, X might say, 'I am feeling a pain in my right knee. If there is anyone here with pain there, come and be prayed for.' Sometimes the non-specificity of these sympathetic pains (or the mention of an ailment in a so-

called 'word of knowledge') can lead to cynicism. If you say to a congregation of a thousand, 'There is someone here with lower back pain,' it would be very surprising if you were wrong. Charles Williams, a member of the Inklings, thought that intensely sympathetic involvement in the physical or mental pain of another could relieve that pain by the pain or illness actually being transferred to the willing, sympathetic recipient. He saw this as a corollary of the biblical injunction, 'Bear one another's burdens' (Gal. 6:2), and called it the 'doctrine of substituted love'. It is best articulated in his book *Descent into Hell* (1937).

70 This is no place for an essay on the undoubtedly awesome power of the mind, but the issue needs a mention. The cleanest and least controversial example is the placebo effect. It is massively under-researched, mainly because most of the funding for medical research comes from drug companies and other commercial organisations, who of course not only have no interest in investigating the placebo effect, but a definite interest in it being ignored. There is a superb discussion in B. H. Lipton, *The Biology of Belief* (Carlsbad, California: Hay House, 2008), pp. 93–114.

Chapter 5: Getting Out of Yourself: An Introduction to Other States of Consciousness

1 W. James, *What Psychical Research Has Accomplished* (1892).

2 A. Huxley, 'Drugs that shape men's minds', *Saturday Evening Post*, 18 October 1958.

3 *Encyclopaedia Britannica* (2006), 'Hypnosis'.

4 A. Gosline, 'Hypnosis really changes your mind', *New Scientist*, 10 September 2004.

5 *Encyclopaedia Britannica* (2006), 'Hypnosis'.

6 In Acts 10, Peter, in Joppa, went onto the roof of a house to pray. He was hungry (a point to which we will return in Chapter 9), and while the food was being prepared 'he fell into a trance. He saw heaven opened and something like a large sheet being let down to

earth by its four corners. It contained all kinds of four-footed animals, as well as reptiles of the earth and birds of the air.' Peter was told to 'kill and eat'. He protested, saying that he would not eat unclean creatures. He is told that they are clean. And so Peter learned that the gospel was for Gentiles as well as for Jews (Acts 10:9–16). Paul, having been converted on the Damascus road, returned to Jerusalem. While he was praying in the temple, he says, 'I fell into a trance and saw the Lord speaking. "Quick . . . Leave Jerusalem immediately, because they will not accept your testimony about me"' (Acts 22:17–18). In 2 Corinthians 12:2–4, Paul relates an apparent out-of-body experience which looks very like a classic shamanic journey to another world. It is discussed further in Chapter 8. There is, of course, Paul's conversion itself (Acts 9:3–9; 22:6–11) and (possibly a reference to some sort of altered state of consciousness) the statement about John of Patmos: 'On the Lord's Day I was in the Spirit, and I heard behind me a loud voice like a trumpet, which said: "Write on a scroll what you see"' (Rev. 1:10–11; see Chapters 6 and 11 for further discussion of John of Patmos).

7 G. Hancock, *Supernatural: Meetings with the Ancient Teachers of Mankind* (London: Arrow, 2006), p. 289.
8 Acts 1:6–11.
9 These phenomena are discussed in detail in Chapter 11.
10 Famous examples include St Francis of Assisi, St Alphonsus Liguouri, St Catherine of Siena, St Francis Xavier, St Ignatius Loyola, St Teresa of Avila, St Thomas Aquinas and the (formally uncanonised) Christina the Astonishing.

Chapter 6: Finding God in a Garden: How Psychoactive Substances can Throw Open the Doors of Perception

1 www.albinoblacksheep.com/flash/badgers.
2 Cited in G. Hancock, *Supernatural: Meetings with the Ancient Teachers of Mankind* (London: Arrow, 2006), pp. 703–4.

3 Cited in ibid., p. 632.

4 Cited in ibid., pp. 631–2.

5 See, for instance, R. G. Wasson, A. Hofmann and C. A. P. Ruck, *The Road to Eleusis: Unveiling the Secret of the Mysteries* (New York: Harcourt Brace Jovanovich, 1978); T. McKenna, *Food of the Gods: The Search for the Original Tree of Knowledge. A Radical History of Plants, Drugs and Human Evolution* (New York: Bantam, 1993); H. P. Foley (ed.), *The Homeric Hymn to Demeter: Translation, Commentary, and Interpretive Essays* (Princeton, NJ: Princeton University Press, 1994); I. Valencic, 'Has the mystery of the Eleusinian Mysteries been solved?', http://www.x-sandra.com/valencic/valencic/ivan.htm, 1994.

6 See D. Merkur, *The Mystery of Manna: The Psychedelic Sacrament of the Bible* (Rochester, VT: Park Street Press, 2000).

7 G. Samorini, 'The oldest representations of hallucinogenic mushrooms in the world', http://www.samorini.net.

8 As we see in Chapter 11.

9 P. Barabe, 'The religion of Iboga and the Bwiti of the Fangs', http://www.ibogaine.desk.nl/barabe.html.

10 D. Pinchbeck, 'Ten years of therapy in one night', *Guardian*, 20 September 2003.

11 Ibid.

12 Reported by J. Horgan, 'Peyote on the brain', *Discover* magazine, February 2003. The other details of the peyote ceremony are taken from Horgan's account.

13 A. Huxley, *The Doors of Perception and Heaven and Hell* (Harmondsworth: Penguin, 1959), pp. 5–7.

14 Hancock, *Supernatural*, pp. 50–77.

15 A question to which we return in Chapter 11.

16 *Rig Veda*, 8.48, trans. W. Doniger (Harmondsworth: Penguin, 1981), pp. 134–5.

17 *Rig Veda*, 10.119, ibid., p. 132.

18 *Rig Veda*, 9.113, ibid., pp. 133–4.

19 Other candidates include cannabis, Syrian rue and the mushroom *Psilocybe cubensis*, an inebriating mushroom that grows in cow

dung and is common in India. For discussion of the *cubensis* thesis, see J. Ott, 'The post-Wasson history of the Soma plant, Eleusis', *Journal of Psychoactive Plants and Compounds*, 1:9.

20 See Chapter 11.

21 See Hancock, *Supernatural*, p. 607.

22 Such as Danny Staples in C. A. P. Ruck and D. Staples, *The World of Classical Myth: Gods and Goddesses, Heroines and Heroes* (Durham, NC: Carolina Academic Press, 1994), p. 26.

23 There are various versions of the Tantalus story, and various reasons (of which the theft of ambrosia is one) for his torture. There are obvious parallels with the story of Prometheus, to which we return in Chapter 11.

24 The idea that any psychoactive substance at all was involved at Delphi has been questioned by many, such as J. Fontenrose, *The Delphic Oracle: Its Responses and Operations, with a Catalogue of Responses* (1978), relying on the apparent intelligibility of the oracle's pronouncements. For a spirited defence of the psychoactive thesis, see H. A. Spiller, J. R. Hale and J. Z. de Boer, 'The Delphic Oracle: A Multidisciplinary Defense of the Gaseous Vent Theory', *Clinical Toxicology*, 40.2, 2000, pp. 189–96.

25 French excavators at the end of the nineteenth century and the beginning of the twentieth century found no evidence of any volcanic fissures through which any gases could have permeated. But excavations at the start of the twenty-first century have questioned this: see W. J. Broad, *The Oracle: The Lost Secrets and Hidden Message of Ancient Delphi* (London: Penguin, 2006); J. de Boer, J. R. Hale and J. Chanton, 'New Evidence for the Geological Origins of the Ancient Delphic Oracle', *Geology*, 29.8, 2001, pp. 707–11.

26 Plutarch, *Moralia*, 437.

27 'Thence for nine days' space I was borne by direful winds over the teeming deep; but on the tenth we set foot on the land of the Lotus-eaters, who eat a flowery food. There we went on shore and drew water, and straightway my comrades took their meal by the swift ships. But when we had tasted food and drink, I sent

forth some of my comrades to go and learn who the men were, who here ate bread upon the earth; two men I chose, sending with them a third as a herald. So they went straightway and mingled with the Lotus-eaters, and the Lotus-eaters did not plan death for my comrades, but gave them of the lotus to taste. And whosoever of them ate of the honey-sweet fruit of the lotus, had no longer any wish to bring back word or to return, but there they were fain to abide among the Lotus-eaters, feeding on the lotus, and forgetful of their homeward way. These men, therefore, I brought back perforce to the ships, weeping, and dragged them beneath the benches and bound them fast in the hollow ships; and I bade the rest of my trusty comrades to embark with speed on the swift ships, lest perchance anyone should eat of the lotus and forget his homeward way. So they went on board straightway and sat down upon the benches, and sitting well in order smote the grey sea with their oars' (*Odyssey*, Book 9, trans. A. T. Murray, Plain Label Books, 1946).

28 Herodotus, *Histories*, 4:75.

29 Euripides, *Bacchae*, trans. R. E. Meagher (Waconda, IL: Bolchazy-Carducci, 1995).

30 A. Huxley, 'Drugs that shape men's minds', *Saturday Evening Post*, 18 October 1958.

31 T. De Quincey, *Confessions of an English Opium Eater* (1821).

32 See too Coleridge's experience on what was almost certainly opium, taken in 1797 while he was staying in a lonely Exmoor farmhouse. '[A]ll the images rose up . . . as *things* . . . without any sensation or consciousness of effect.' He was famously woken by the 'person on business from Porlock'. He retained just a few lines of what he had heard, and a few images, but they became *Kubla Khan*.

33 'Entheogen' is a term coined in 1979 by a group of ethnobotanists and scholars of mythology. It means 'generator of a god within'. It has become popular amongst those who think that 'hallucinogen' implies some sort of pathological aetiology to the visions seen with drugs. Terence McKenna disliked the term 'entheogen',

thinking that it bore too much theological baggage and implied too pejoratively a connection between religious experiences and experiences on psychoactive drugs. I agree with him.

34 T. F. Leary, *Flashbacks, An Autobiography* (Los Angeles: J. P. Tarcher, 1983), p. 375.

35 http://leda.lycaeum.org/?ID=9293, taken from the internet newsgroup alt.drugs and related newsgroups.

36 Ibid.

37 Cited in D. Tymoczko, 'The Nitrous Oxide Philosopher', *The Atlantic Monthly*, 27:5, 1996, pp. 93–101.

38 And cannabis may have been important in central Asian shamanism. Mummies dated to 1000 BC, discovered in Xinjiang province, had sacks of marijuana buried close to them. It has been concluded (mainly on the evidence of the cannabis itself) that the bodies were those of shamans who used cannabis in their rituals.

39 Gen. 1:11–12 reads (in the King James Version), 'And God said, Let the earth bring forth grass, the herb yielding seed, and the fruit tree yielding fruit after his kind, whose seed is in itself, upon the earth: and it was so. And the earth brought forth grass, and herb yielding seed after his kind, and the tree yielding fruit, whose seed was in itself, after his kind: and God saw that it was good.' Gen. 3:18 reads, 'Thorns also and thistles shall [the ground] bring forth to thee; and thou shalt eat the herb of the field.' Prov. 15:17 reads, 'Better is a dinner of herbs where love is, than a stalled ox and hatred therewith.' Ps. 104:14 reads, 'He causeth the grass to grow for the cattle, and herb for the service of man: that he may bring forth food out of the earth.'

40 Exod. 30:28. The word normally translated 'calamus' is the Hebrew word *kaneh-bos*. The Polish etymologist Sara Benetowa (Sula Benet) maintained that this was a reference to cannabis; see *Tracing One Word Through Different Languages* (1936) (reprinted in *The Book of Grass*, 1967). See too A. Kaplan, 'The Living Torah', 1981, http://bible.ort.org/books/pentd2.asp.

41 See, e.g., http://www.freeanointing.org/cannabis_in_the_holy_oil.htm.

42 Gen. 3:7. The whole of the relevant passage is Gen. 2:8 – 3:24: 'And the Lord God planted a garden in Eden, in the east; and there he put the man whom he had formed. Out of the ground the Lord God made to grow every tree that is pleasant to the sight and good for food, the tree of life also in the midst of the garden, and the tree of the knowledge of good and evil . . . And the Lord God commanded the man, "You may freely eat of every tree of the garden; but of the tree of the knowledge of good and evil you shall not eat, for in the day that you eat of it you shall die." . . . Now the serpent was more crafty than any other wild animal that the Lord God had made. He said to the woman, "Did God say, 'You shall not eat from any tree in the garden'?" The woman said to the serpent, "We may eat of the fruit of the trees in the garden; but God said, 'You shall not eat of the fruit of the tree that is in the middle of the garden, nor shall you touch it, or you shall die.'" But the serpent said to the woman, "You will not die; for God knows that when you eat of it your eyes will be opened, and you will be like God, knowing good and evil." So when the woman saw that the tree was good for food, and that it was a delight to the eyes, and that the tree was to be desired to make one wise, she took of its fruit and ate; and she also gave some to her husband, who was with her, and he ate. Then the eyes of both were opened, and they knew that they were naked; and they sewed fig leaves together and made loincloths for themselves. They heard the sound of the Lord God walking in the garden at the time of the evening breeze, and the man and his wife hid themselves from the presence of the Lord God among the trees of the garden. But the Lord God called to the man, and said to him, "Where are you?" He said, "I heard the sound of you in the garden, and I was afraid, because I was naked; and I hid myself." He said, "Who told you that you were naked? Have you eaten from the tree of which I commanded you not to eat?" The man said, "The woman whom you gave to be with me, she gave me fruit from the tree, and I ate." Then the Lord God said to the woman, "What is this that you have done?" The woman

said, "The serpent tricked me, and I ate." The Lord God said to the serpent, "Because you have done this, cursed are you among all animals and among all wild creatures; upon your belly you shall go, and dust you shall eat all the days of your life. I will put enmity between you and the woman, and between your offspring and hers; he will strike your head, and you will strike his heel." To the woman he said, "I will greatly increase your pangs in childbearing; in pain you shall bring forth children, yet your desire shall be for your husband, and he shall rule over you." And to the man he said, "Because you have listened to the voice of your wife, and have eaten of the tree about which I commanded you, 'You shall not eat of it,' cursed is the ground because of you; in toil you shall eat of it all the days of your life; thorns and thistles it shall bring forth for you; and you shall eat the plants of the field. By the sweat of your face you shall eat bread until you return to the ground, for out of it you were taken; you are dust, and to dust you shall return." Then the Lord God said, "See, the man has become like one of us, knowing good and evil; and now, he might reach out his hand and take also from the tree of life, and eat, and live forever" – therefore the Lord God sent him forth from the garden of Eden, to till the ground from which he was taken. He drove out the man; and at the east of the garden of Eden he placed the cherubim, and a sword flaming and turning to guard the way to the tree of life.'

43 R. G. Wasson, *Soma: Divine Mushroom of Immortality* (The Hague: Mouton, 1968), p. 221.

44 There are some real and underplayed difficulties in interpreting the Eden story. I have discussed them in detail in *The Selfless Gene: Living with God and Darwin* (London: Hodder & Stoughton, 2009). But the notion that Yahweh is secretly pleased with the illegitimate plucking of the fruit is exegetically impossible. The whole story of the Old Testament is an account of Yahweh's displeasure with the presumption that both caused and resulted from the taking of the fruit, and his patience and forbearance with his presumptuous, disobedient people. The whole story

of the New Testament is an account of Yahweh's supremely costly unravelling of the consequences of the fruit-taking.

45 There is no doubt about Allegro's credentials as a biblical scholar. He was involved, for instance, in the translation of the Dead Sea Scrolls.

46 J. Allegro, 'Sacred Mushroom', *Sunday Mirror*, 12 April 1970, p. 10.

47 A good example is the curious juxtaposition of the mushroom and the alleged spaceship in one of Hildegard of Bingen's illustrations, discussed below.

48 Things have moved on little since Erwin Panofsky wrote to Wasson in 1952, '[T]he plant in this fresco has nothing whatever to do with mushrooms ... and the similarity with *Amanita muscaria* is purely fortuitous. The Plaincourault fresco is only one example – and since the style is provincial, a particularly deceptive one – of a conventionalized tree type, prevalent in Romanesque and early Gothic art, which art historians actually refer to as a "mushroom tree" or in German *Pilzbaum*. It comes about by the gradual schematization of the impressionistically rendered Italian pine tree in Roman and early Christian painting, and there are hundreds of instances exemplifying this development – unknown of course to mycologists ... What the mycologists have overlooked is that the mediaeval artists hardly ever worked from nature but from classical prototypes which in the course of repeated copying became quite unrecognizable' (in Wasson, *Soma*, pp. 179–80).

49 Rev. 2:7 reads, 'To everyone who conquers, I will give permission to eat from the tree of life that is in the paradise of God.' Rev. 22:1–2 reads, 'Then he showed me the river of the water of life, bright as crystal, flowing from the throne of God and of the Lamb through the middle of the street of the city; also, on either side of the river, the tree of life with its twelve kinds of fruit, yielding its fruit each month; and the leaves of the tree were for the healing of the nations.' Rev. 22:14 reads, 'Blessed are those who wash their robes, that they may have the right to the tree of life and that they may enter the city by the gates.'

50 Ezek. 2:9 – 3:4.

51 C. Heinrich, *Strange Fruit: Alchemy, Religion and Magical Foods, a Speculative History* (London: Bloomsbury, 1995), pp. 100, 129.

52 Rev. 10:8–11.

53 A similar case can be made from Jer. 15:16 ('Your words were found, and I ate them, and your words became to me a joy and the delight of my heart') and, much more dubiously, from Ps. 19:10 (the ordinances of the Lord are 'sweeter . . . than honey, and drippings of the honeycomb') and Ps. 119:103 ('How sweet are your words to my taste, sweeter than honey to my mouth!').

54 Huxley, 'Drugs that shape men's minds', ibid.

55 Pinchbeck, 'Ten years of therapy in one night', ibid.

56 Neither the recipients nor the researchers knew who had received what.

57 J. Malgren, 'Tune in, turn on, get well?', *St Petersburg Times*, 27 November 1994, http://www.csp.org/practices/entheogens/docs/young-good_friday.html.

58 R. Doblin, 'Pahnke's Good Friday Experiment: A long-term follow-up and methodological critique', *Journal of Transpersonal Psychology*, 23:1, 1991, pp. 1–28.

59 A group at Johns Hopkins University recently looked at the effect of psilocybin in 36 hallucinogen-naive adults, 53 per cent of whom declared some sort of affiliation with a religious institution, and all of whom had at least intermittent participation in religious or spiritual activities. Most of the subjects rated the resultant experience as amongst the top five most spiritually significant experiences in their lives, and two months later nearly 80 per cent reported increased well-being or satisfaction. Fourteen months later, most of the subjects continued to rate the experience the most or one of the five most personally meaningful and spiritually significant experiences of their lives; see R. R. Griffiths, W. A. Richards, M. W. Johnson, U. McCann and R. Jesse, 'Psilocybin can occasion mystical-type experiences having substantial and sustained personal meaning and spiritual significance', *Journal of Psychopharmocology*, 187(3), 2006,

pp. 268–83; and R. R. Griffiths et al., 'Mystical type experiences occasioned by psilocybin mediate the attribution of personal meaning and spiritual significance 14 months later', *Journal of Psychopharmacology*, 22(6), 2008, pp. 621–32.

60 Actually 22 per cent; see Griffiths et al., 'Psilocybin can occasion mystical-type experiences'.

61 A recent paper by Sara Lewis emphasised the spiritual and psychological crises faced by Western users of ayahuasca. S. E. Lewis, 'Ayahuasca and Spiritual Crisis: Liminality as Space for Personal Growth', *Anthropology of Consciousness*, 19:2, 2008, pp. 109–33.

62 J. Horgan, *Rational Mysticism* (Boston/New York: Houghton Mifflin, 2003), p. 211.

63 K. L. R. Jansen, 'Ketamine, Near-Birth and Near-Death Experiences', in *Ketamine: Dreams and Realities* (Florida: Multidisciplinary Association for Psychedelic Studies, 2001), pp. 92–166 (original emphasis).

64 Actually 47 per cent had these encounters.

65 The story is told in Strassman's book *DMT: The Spirit Molecule: A Doctor's Revolutionary Research into the Biology of Near-Death and Mystical Experiences* (New York: Park Street Press, 2001).

66 Discussed in Chapter 2.

67 In N. Saunders, 'The agony and ecstasy of God's path', *Guardian*, 29 July 1995.

68 Ibid.

69 Ibid.

70 Matt. 26:26–8. See also Mark 14:22–4 and Luke 22:17–20.

71 Huxley, 'Drugs that shape men's minds', ibid.

Chapter 7: Finding God in the Bedroom: The Sexuality of Spirituality and the Spirituality of Sexuality

1 D. G. White, *Tantra in Practice* (Princeton: Princeton University Press, 2000), p. 7.

2 We will see in Chapter 7 that Newberg posits an evolutionary connection between the mystical experience in general and the experience of orgasm in particular.

3 Song of Songs 4:5.

4 See, for instance, Lev. 18.

5 St John of the Cross, *Spiritual Canticle*, 22, 27, 28, 38–40, trans. David Lewis (1889). All of it is well worth a read. Some further examples: 'Beneath the apple-tree/There were you betrothed;/There I gave you My hand/And you were redeemed/Where your mother was corrupted/Our bed is of flowers/By dens of lions encompassed/Hung with purple/Made in peace/And crowned with a thousand shields of gold. . ./My soul is occupied/And all my substance in His service/Now I guard no flock/Nor have I any other employment/My sole occupation is love. . ./If, then, on the common land/I am no longer seen or found/You will say that I am lost/That, being enamored, I lost myself; and yet was found./Of emeralds, and of flowers/In the early morning gathered, We will make the garlands/Flowering in Your love/And bound together with one hair of my head./By that one hair/You have observed fluttering on my neck/And on my neck regarded, You were captivated/And wounded by one of my eyes./When You regarded me/Your eyes imprinted in me Your grace/For this You loved me again/And thereby my eyes merited/To adore what in You they saw. . ./The little white dove/Has returned to the ark with the bough; And now the turtle-dove/Its desired mate/On the green banks has found./In solitude she lived, And in solitude built her nest/And in solitude, alone/Has the Beloved guided her/In solitude also wounded with love./Let us rejoice, O my Beloved!/Let

us go forth to see ourselves in Your beauty/To the mountain and the hill/Where the pure water flows/Let us enter into the heart of the thicket' (ibid., 24, 26, 29–37).

6 Celibacy in the Old Testament is an aberration. It was sometimes adopted in extreme circumstances to make a prophetic point. See, e.g., Jer. 16.

7 Predictably, some of the most strident denunciations of the misuse of sex come from the monastic world, but also (perhaps it is an indication of prurient interest) some of the most accurately recorded observations about human sexual practice also come from there. Sandwiched between Hildegard of Bingen's terse sermons about the dangers of self-pollution and other unnatural sexual activity comes this breathtakingly candid description of female orgasm: 'When a woman is making love with a man, a sense of heat in her brain, which brings with it sensual delight, communicates the taste of that delight during the act and summons forth the emission of the man's seed. And when the seed has fallen into its place, that vehement heat descending from her brain draws the seed to itself and holds it, and soon the woman's sexual organs contract, and all the parts that are ready to open up during the time of menstruation now close, in the same way as a strong man can hold something enclosed in his fist' (trans. S. Flanagan, *Hildegard of Bingen, 1098–1179: A Visionary Life*, London: Routledge, 1998, p. 97). It is strange stuff from a celibate abbess. If you want to link the interest and the insight to her migrainous episodes there is nothing to stop you apart from a complete absence of any evidence justifying it. Interestingly Hildegard, despite her apparently encyclopaedic knowledge of human sexuality, falls well behind many of the other, apparently more sexually naive mystics, in using erotic language to describe her own mystical experiences.

8 See C. Foster, *The Christmas Mystery* (Milton Keynes/Colorado Springs/Hyderabad: Authentic, 2007).

9 *Symposium* 211:b-c, trans. R. Waterfield (Oxford: Oxford World Classics, 1998).

10 His influence continues to spread widely today, helped along mightily by C. S. Lewis, who had a blind spot when it came to Plato.

Chapter 8: Finding God in the Intensive Care Unit: Near-death and Other Out-of-body Experiences

1 http://www.near-death.com/experiences/research16.html.

2 2 Cor. 12:2–4.

3 'Out-of-body experience' is the currently respectable label. 'Astral projection' – by which is meant the same thing – is out of vogue: it carries too much dubious spiritualist baggage.

4 S. Blackmore, 'A postal survey of OBEs and other experiences', *Journal of the Society for Psychical Research*, 52, 1984, pp. 225–44. She found that 12 per cent of Bristol residents had had an OBE. Typically they were resting or lying down when they suddenly felt that they had left their bodies. Usually the experience lasted for one or two minutes.

5 Magritte described his own paintings as 'the representations of half sleep'.

6 The scientific pioneer of sensory deprivation by way of flotation tanks was John Lilly. See J. C. Lilly and E. J. Gold, *Tanks for the Memories: Flotation Tank Talks* (Gateways, 2000); J. C. Lilly, *The Deep Self: Profound Relaxation and the Tank Isolation Technique* (Warner Books, 1981).

7 O. Blanke, S. Ortigue, T. Landis and M. Seeck, 'Out of body experience and autoscopy of neurological origin', *Brain*, 127:2, 2004, pp. 243–8.

8 See Chapter 2.

9 H. H. Ehrsson, 'The experimental induction of out-of-body-experiences', *Science*, 317, 2007, p. 1048.

10 R. Moody, *Life after Life* (Covinda, Gal: Mockingbird, 1975); F. Schoonmaker, 'Denver cardiologist discloses findings after 18 years of near-death research', *Anabiosis*, 1, 1979, pp. 1–12; K. Ring,

Life at Death: A Scientific Investigation of the Near-Death Experience (New York: Coward, McCann and Geoghegan, 1980).

11 P. Van Lommel, R. Van Wees, V. Meyers and I. Elfferich, 'Near-death experience in survivors of cardiac arrest: a prospective study in the Netherlands', *The Lancet*, 358, 2001, pp. 2039–45.

12 A. J. Ayer, 'What I saw when I was dead', *National Review*, 14 October 1988.

13 It is commonly said that something akin to classic NDEs is seen in the eighth-century AD Tibetan *Book of the Dead*, a handbook that tells the dead what to expect between death and rebirth. I don't find the parallels convincing myself.

14 *Republic*, trans. B. Jowett (London: Courier Dover, 2000).

15 Van Lommel et al., 'Near-death experience in survivors of cardiac arrest'.

16 The story is told in M. Sabom, *Light and Death: One Doctor's Fascinating Account of Near-Death Experiences* (Grand Rapids, MI: Zondervan, 1998), and summarised at http://near-death.com/experiences/evidence01.html.

17 Ring, *Life at Death*, ibid.

18 Ring, quoted in K. Jansen, 'Ketamine: Near Death and Near Birth Experiences', http://lila.info/document_view.phtml?document_id=91.

19 K. L. R. Jansen, 'Ketamine, Near-Birth and Near-Death Experiences', in *Ketamine: Dreams and Realities* (Florida: Multidisciplinary Association for Psychedelic Studies, 2001).

20 The imbalance is corrected by Jansen, ibid.

21 C. Zaleski, *Otherworld Journeys: Accounts of Near Death Experience in Mediaeval and Modern Times* (New York: Oxford University Press, 1989).

22 Moody, *Life after Life*, cited in Jansen, 'Ketamine, Near-Birth and Near-Death Experiences', in *Ketamine*.

23 J. Morse, P. Castillo, D. Venecia, J. Milstein and D. C. Tyler, 'Childhood near-death experiences', *American Journal of Diseases of Children*, 140, 1986, pp. 1110–14, cited in S. Blackmore, 'Near-

Death Experiences: In or Out of the Body?', *Skeptical Inquirer*, 16, 1991, pp. 34–45.

24 K. Ring and S. Cooper, *Mindsight: Near-Death and Out-of-Body Experiences in the Blind* (Institute of Transpersonal Psychology, 1999).

25 http://www.near-death.com/evidence.html at 31.

26 'Coping with life after near-death experience', *New York Times*, http://neardeath.home.comcast.net/nde/001_pages/69.htm.

27 60 per cent of the subjects who had had NDEs reported that they had had the REM stage of sleep while they were still awake. The corresponding figure for those who had not had an NDE was 25 per cent. K. R. Nelson, M. Mattingly, S. A. Lee and F. A. Schmitt, 'Does the arousal system contribute to near death experiences?', *Neurology*, 66, 2006, pp. 1003–9.

28 Ring, *Life at Death*, ibid.

29 Van Lommel et al., 'Near-death experience in survivors of cardiac arrest'. Short-term memory correlated independently with reporting of the NDE. The authors commented, at p. 2043: 'Good short-term memory seems to be essential for remembering NDE. Patients with memory defects after prolonged resuscitation reported fewer experiences than other patients in our study. Forgetting or repressing such experiences in the first days after CPR was unlikely to have occurred in the remaining patients, because no relation was found between frequency of NDE and date of first interview.'

30 See, for instance, W. Barrett, *Death-Bed Visions* (London: Methuen, 1926).

31 Dr Carla Wills-Brandon contends that only about 10 per cent of people are conscious shortly before their death, but of this 10 per cent, 50–67 per cent have deathbed visions; see 'Death-bed visions: Dr Carla Wills-Brandon's research', http://www.near-death.com/deathbed.html.

32 C. T. Tart, 'A psychophysiological study of out-of-body experiences in a selected subject', *Journal of the Society for Psychical Research*, 62, 1978, pp. 3–27.

33 See 'World's largest-ever study of near-death experiences' (2008), http://www.soton.ac.uk/mediacentre/news/2008/sep/08_165.shtml.

34 For instance, T. Lempert, M. Bauer and D. Schmidt, 'Syncope and near-death experience', *The Lancet*, 344, 1994, pp. 829–30.

35 L. Appleby, 'Near-death experience: analogous to other stress induced physiological phenomena', *British Medical Journal*, 298, 1989, pp. 976–7.

36 See S. Grof and I. Halifax, *The Human Encounter with Death* (London: Souvenir Press, 1977); C. Sagan, *Broca's Brain* (New York: Random House, 1979).

37 S. Blackmore, 'Birth and the OBE: An unhelpful analogy', *Journal of the American Society for Psychical Research*, 77, 1982, pp. 229–38. The final nail would be a similar study relating specifically to NDEs.

38 This is what is known as disinhibition – the reduction of the braking, regulating function of the brain.

39 J. D. Cowan, 'Spontaneous symmetry breaking in large-scale nervous activity', *International Journal of Quantum Chemistry*, 22, 1982, pp. 1059–82, cited in Blackmore, 'Near-Death Experiences', *Skeptical Inquirer*, 16, pp. 34–45.

40 Blackmore, ibid.

41 See Chapter 2.

42 Blackmore, 'Near-Death Experiences', ibid.

43 R. Kinseher, *Geborgen in Liebe und Licht: Gemeinsame Ursache von Intuition, Déjà-vu-, Schutzengel- und Nahtod-Erlebnissen* (2006); R. Kinseher, *Verborgene Wurzeln des Glücks: Selbstbeobachtbare Gehirnfunktion* (Books on Demand Gmbh, 2008).

44 J. C. Saavedra-Aguilar and Gomez-Jeria, 'A neurobiological model for near-death experiences', *Journal of Near Death Studies*, 7, 1989, pp. 205–22, cited in Blackmore, 'Near-Death Experiences'.

45 Blackmore, ibid.

46 B. Greyson, 'Post traumatic stress symptoms following near-death experiences', *American Journal of Orthopsychiatry*, 71, 2001, pp. 368–73.

47 See Chapter 6.

48 It is also an excellent analgesic.

49 K. L. R. Jansen, 'Using ketamine to induce the near-death experience: mechanism of action and therapeutic potential', *Yearbook for Ethnomedicine and the Study of Consciousness*, 4, ed. C. Ratsch and J. R. Baker (Berlin: VWB, 1995), pp. 55–81. See too his comprehensive review of the relationship between ketamine and NDEs in K. L. R. Jansen, 'Ketamine, Near-Birth and Near-Death Experiences', in *Ketamine: Dreams and Realities* (Florida: Multidisciplinary Association for Psychedelic Studies, 2001), pp. 92–166.

50 Ibid.

Chapter 9: Other Portals

1 Cited in J. De Marchi, *The True Story of Fatima* (St Paul, Minnesota: Catechetical Guild Entertainment Society, 1952), p. 143.

2 These are broadly the objections of Chris Rutkoswki of the University of Manitoba; see http://www.holman.net/ufo/archives/miscnewfiles/rutkowski/ persinger.

3 'Origin of the Book of Mormon', in *The Book of Mormon*, numerous editions.

4 R. P. Bentall and P. D. Slade, 'Reliability of a scale measuring disposition towards hallucination: a brief report', *Personality and Individual Differences*, 6, 1985, pp. 527, 529.

5 W. D. Rees, 'The hallucinations of widowhood', *British Medical Journal*, 4, 1971, pp. 37–41. Although hallucinations are generally more common in women than men, more men than women had bereavement hallucinations in Rees's study: 50 per cent men; 45.8 per cent women.

6 C. Foster, *The Jesus Inquest* (Oxford: Monarch, 2006), ch. 6.

7 M. Winkleman, 'Trance States: A Theoretical Model and Cross-Cultural Analysis', *Ethos*, 14, 1986, pp. 174–203; M. Winkleman, *Shamanism: The Neural Ecology of Consciousness and Healing* (Westport: Bergin & Garvey, 2000).

8 See too G. W. Dennis, 'The Use of Water as a Medium for Altered

States of Consciousness in Early Jewish Mysticism: A Cross-Disciplinary Analysis', *Anthropology of Consciousness*, 19(1), 2008, pp. 84–106.

9 H. Benson, J. W. Lehmann, M. S. Malhotra, R. F. Goldman and J. E. Hopkins, 'Body temperature changes during the practice of Tum-mo yoga', letter to *Nature* magazine, 21 January 1982, *Nature* 295, pp. 234–6; W. J. Cromie, 'Research: Meditation changes temperatures: Mind controls body in extreme experiments', Cambridge, MA: *Harvard University Gazette*, 18 April 2002.

10 These effects were certainly known in antiquity, and it is suggested that some structures (e.g. Neolithic passage tombs, including Newgrange) might have been specifically designed to trap sounds as 'standing waves' – amplifying the volume and making it appear as if it comes from everywhere at once; see S. E. Hale and D. Campbell, 'Sacred Space, Sacred Sound', London: *Quest*, 2007, p. 62, summarising the work of Aaron Watson.

11 The word literally means 'the physical basis'.

12 Dalai Lama, *Freedom in Exile* (New York: HarperOne, 1991).

13 C. S. Lewis, *Surprised by Joy* (London: Fount, 1983), pp. 18–19.

Chapter 10: Turning On and Tuning In: Brains as Antennae

1 See Chapter 2.

2 http://www.near-death.com/evidence.html, at 13.

3 Absolute identity with everything else in the universe, and experiential knowledge of the fallacy of dualism.

4 I don't deal here with the (presumably related) question of whether brains can transmit as well as receive. I will simply record that I'd be very unhappy about any scientific worldview or theology that couldn't accommodate the notion that brains can broadcast.

5 Presumably a reference to the beta waves of the normal waking brain, seen on EEG.

6 Paul Roussow, 1994, in a drugs newsgroup archived at http://leda.lycaeum.org/?ID =9293.

7 R. Strassman, *DMT: The Spirit-Molecule: A Doctor's Revolutionary Research into the Biology of Near-Death and Mystical Experiences* (New York: Park Street Press, 2001), cited by G. Hancock, *Supernatural: Meetings with the Ancient Teachers of Mankind* (London: Arrow, 2006), pp. 357–8.

8 Roussow, http://leda.lycaeum.org/?ID =9293.

9 All from W. James, 'The Subjective Effects of Nitrous Oxide', *Mind*, vol. 7, 1882.

10 C. Foster, *The Selfless Gene: Living with God and Darwin* (London: Hodder & Stoughton, 2009).

11 And indeed we return to the second question in this and the following chapter.

12 A. Newberg and M. R. Waldman, *Born to Believe* (New York: Free Press, 2006), p. xxi.

13 Ibid., p. xvii.

14 'Cognitive imperative', as far as I can see, is an expression coined by Newberg.

15 Newberg and D'Aquili, *Why God Won't Go Away*, p. 60.

16 Letter to Reginald Berkeley, cited in Edward Marsh, *Memoir on the Life of Rupert Brooke* (1918).

17 They contend that the parietal lobe is the seat of the brain's main causal and binary operators, and of language, and accordingly that 'whoever the first myth-makers actually were, they were likely set apart neurologically from the rest of creation by the presence and function of a well-developed parietal lobe' (Newberg and D'Aquili, *Why God Won't Go Away*, pp. 64–72).

18 C. Levi-Strauss, *Structural Anthropology*, trans. C. Jacobsen and B. Grundfest Schoepf (Harmondsworth: Penguin, 1968).

19 L. O. McKinney, *Neurotheology: Virtual Religion in the 21st Century* (American Institute for Mindfulness, 1994).

20 See Chapter 2.

21 And in 'Tintern Abbey' he described an experience akin to Absolute Unitary Being that would make any advanced Indian yogi say, 'Yes, he's been there too': 'that serene and blessed mood,/In which the affections gently lead us on,/Until the breath of this

corporeal frame/And even the motion of our human blood/Almost suspended, we are laid asleep/In body, and become a living soul:/While with an eye made quiet by the power/Of harmony, and the deep power of joy,/We see into the life of things.'

22 Wordsworth himself, of course, believed something very different from Freud, and would be outraged by the association. In Wordsworth's mind the title refers (in Iain McGilchrist's words) 'not [to] the genesis of a self-serving myth, but [to] the fact that our familiar ways of being-in-the-world in adulthood screen out the other frequencies on your dial' (personal communication, 2009). As the rest of this chapter indicates, I am in this respect a passionate Wordsworthian and a passionate opponent of Freud.

23 See S. E. Guthrie (1993) *Faces in the clouds: A new theory of religion*, New York: Oxford University Press. Discussed in J. Barrett, *Why Would Anyone Believe in God?* (Lanham, MD: AltaMira, 2004), p. 31.

24 O. Petrovich, 'Understanding of non-natural causality in children and adults: A case against artificialism', *Psyche ed Geloof*, 8, 1997, pp. 151–65; O. Petrovich, 'Preschool children's understanding of the dichotomy between the natural and the artificial', *Psychological Reports*, 84, 1999, pp. 3–27, cited in Barrett, ibid., pp. 84–5.

25 See E. M. Evans, 'Cognitive and contextual factors in the emergence of diverse belief systems: Creation versus evolution', *Cognitive Psychology*, 42, 2001, pp. 217–66, cited in Barrett, ibid., p. 85. Also D. Kelemen, 'Beliefs about purpose: on the origins of teleological thought', in M. Corballis and S. Lea (eds), *The Descent of Mind: Psychological Perspectives on Hominid Evolution* (Oxford: Oxford University Press, 1999), pp. 278–94; D. Kelemen, 'Why are rocks pointy? Children's preference for teleological explanations of the natural world', *Developmental Psychology*, 35, 1999, pp. 1440–53, cited in Barrett, ibid., p. 85, noting that young children have a strong tendency to attribute *purpose* to whatever animate and inanimate things do. If a rock rolls down a slope, it does so, to a three-year-old, with a purpose.

26 For further discussion of this experiment, see Barrett, ibid., pp. 78–9.

27 R. A. Richert and J. L. Barrett, 'Do you see what I see? Young children's assumptions about God's perceptual abilities', *International Journal for the Psychology of Religion*, cited in Barrett, ibid., p. 81.

28 There is another very strange thing about children's brains. In primates, and only primates, including humans, there are some curious cells in the frontoinsular cortex, called spindle cells. In humans they appear at around four months of age, and they expand in size and multiply for the first three years. They are linked to the acquisition of a moral sense. They light up when their owner sees somebody being deceived or treated unfairly; see Newberg and Waldman, *Born to Believe*, p. 118.

29 Since human agency is one of the most common forms of agency that we see, and 'human' is probably the default candidate when one is trying to identify an agent, I suppose that Guthrie and Piaget might use HADD – an idea that we meet in the main text very shortly – to try to recover some of their anthropomorphism thesis. But this is semantic stuff, and doesn't get us anywhere.

30 Barrett, ibid., p. 31.

31 Ibid., pp. 42–3.

32 Ibid.

33 Ibid., p. 87.

34 The issue is discussed in more detail in Foster, *The Selfless Gene*, ibid.

Chapter 11: Religious Experience and the Origin of Religion

1 See A. Newberg, *God and the Brain* (audiotape, 2007). The phenomenon is well documented. Part of it is expressed in the well-known saying that 'neurones that fire together, wire together'.

2 Cited in M. Alper, *The God part of the brain* (Naperville, IL: Sourcebooks, 2006), p. 77.

3 Ibid.

4 J. Barrett, *Why Would Anyone Believe in God?* (Lanham, MD: AltaMira, 2004), p. 112. I have paraphrased shamelessly.

5 Ibid., p. 124.

6 I. Tattersall, *The Monkey in the Mirror: Essays on the Science of What Makes Us Human* (Oxford/New York: Oxford University Press, 2002), p. 141.

7 C. Foster, *The Selfless Gene: Living with God and Darwin* (London: Hodder & Stoughton, 2009).

8 For instance Colin Renfrew and Ian Tattersall.

9 Foster, *The Selfless Gene*, ibid.

10 The last European paintings are about 12,000 years ago.

11 There are barely a hundred human figures in all three hundred European caves where Upper Palaeolithic paintings are known.

12 See G. Hancock, *Supernatural: Meetings with the Ancient Teachers of Mankind* (London: Arrow, 2006), pp. 160–209.

13 For a discussion and an evaluation of these hypotheses, see ibid., pp. 174–90.

14 Ibid., p. 209.

15 M. Eliade, *Shamanism: Archaic Techniques of Ecstasy* (New York: Routledge and Kegan Paul, 1972).

16 J. Halifax, *Shaman: The Wounded Healer* (New York: Crossroad, 1982).

17 These examples are given in Hancock, *Supernatural*, pp. 333–7.

18 E. A. Peers, *Studies of the Spanish Mystics* (London: SPCK, 1951).

19 Shamans use many techniques to burst through into the spirit world – the techniques described by the Romanian scholar Mircea Eliade as 'archaic techniques of ecstasy'; see Eliade, *Shamanism*.

20 Hancock, *Supernatural*, p. 226, records the comment of a South African practitioner: 'Trance medicine really hurts!'

21 Another, particularly eloquent, version of the thesis that hallucinogens triggered the transformation of our stolid, unimaginative ancestors into vibrantly symbolising modern humanity is in T. McKenna, *Food of the Gods: The Search for the Original Tree*

of Knowledge. A Radical History of Plants, Drugs and Human Evolution (New York: Bantam, 1993).

22 In Chapters 5, 7 and 9.

23 D. Lewis-Williams and D. Pearce, *Inside the Neolithic Mind: Consciousness, Cosmos and the Realm of the Gods* (London: Thames and Hudson, 2005), p. 232.

24 Ibid., p. 252.

25 Ibid., p. 249.

26 Hancock, *Supernatural*, ibid., p. 258.

27 Ibid., p. 452.

28 Conversation with John Horgan, cited in *Rational Mysticism* (Boston/New York: Houghton Mifflin, 2003), p. 183. See too T. McKenna, *The Archaic Revival* (San Francisco: Harper-SanFrancisco, 1991); T. McKenna, *True Hallucinations: Being an Account of the Author's Extraordinary Adventures in the Devil's Paradise* (San Francisco: HarperSanFrancisco, 1993).

29 Eccles. 3:11.

30 See Hancock, *Supernatural*, ibid., pp. 505–6.

31 He points to work that suggests that 'junk' DNA perfectly obeys Zipf's law (which states that for any naturally occurring language, the frequency of occurrence of any word is reciprocally related to its rank in the table of frequencies. Thus the most frequent word will occur about twice as often as the second most frequent word, which in turn will occur twice as often as the fourth most frequent word, and so on). Coding DNA does not. Ibid., pp. 588–93.

32 Ibid., p. 569.

33 Exod. 19:16–18.

34 Exod. 34:29.

35 Exod. 34:30.

36 Exod. 34:33–5.

37 See C. Foster, *Tracking the Ark of the Covenant* (Oxford: Monarch, 2006).

38 Hancock, *Supernatural*, ibid., pp. 440–1.

39 2 Kgs 2:11.

40 Ezek. 1:4–22.

41 Gen. 6:1–4.

42 Cited in Hancock, ibid. *Supernatural*, ibid., p. 626.

43 Rev. 1:10.

44 Rev. 1:13–16: an example not in fact deployed by Graham Hancock, but of a piece with the other instances he uses.

45 Hancock, *Supernatural*, ibid. pp. 604–5.

46 A. Newberg and M. R. Waldman, *Born to Believe* (New York: Free Press, 2006), p. 38.

47 William James agrees with Newberg: 'Suppose, for instance, that you are climbing a mountain, and have worked yourself into a position from which the only escape is by a terrible leap. Have faith that you can successfully make it, and your feet are nerved to its accomplishment. But mistrust yourself, and think of all the sweet things you have heard the scientists say of *maybes*, and you will hesitate so long that, at last, all unstrung and trembling, and launching yourself in a moment of despair, you roll in the abyss' ('Is life worth living?', in *The Will to Believe*, 1897).

48 See Foster, ibid., *The Selfless Gene*.

49 Cited in Hancock, *Supernatural*, p. 252.

50 Cited in ibid., p. 253.

51 Nor is there any reason to agree with the more modest thesis of R. Gordon Wasson that the religious impulse originated as a reaction to the experiences engendered by the (presumably accidental) ingestion of hallucinogenic plants.

52 Attempts to explain away religion either as something that confers a selective advantage or as a by-product of something else (typically symbolic thought) that confers a selective advantage have a long and intellectually disreputable history. They are outside the scope of this book. I have dealt with them in detail in *The Selfless Gene*, ibid.

Chapter 12: Angels or Demons?
The Suppression of Religious Experience

1 In *The Religions of Mongolia* (London: Routledge and Kegan Paul, 1980), p. 36.

2 He appealed against the conviction, and the Supreme Court found that the law under which he had been convicted was unconstitutional.

3 I am emphatically not saying that there are no justifications for legal control of hallucinogenic drugs: there are. The issues are complex, and well outside the scope of this book. Worldwide, there are many restrictions on the use of psychoactive drugs. Although the Native American Church has secured an exemption in relation to peyote from the drug laws that bind everyone else, most of the traditional and new hallucinogenic substances remain illegal in the US and in many other places. Even the famously liberal Netherlands has made the cultivation and use of psilocybin mushrooms unlawful. In the UK it was once hard to pass a shop in Camden Market without seeing magic mushrooms for sale. It is now a criminal offence to possess or sell them.

4 E. V. Pike and F. Cowan, 'Mushroom ritual versus Christianity', *Practical Anthropology*, 6(4), 1959, pp. 145–50.

5 What he asked was, 'What kind of matter is the alien abduction phenomenon? . . . It seems to belong to that class of phenomenon, not even generally accepted as existing by mainstream Western science, that seem not to be *of* this visible, known, material universe and yet appear to manifest *in* it. These are phenomena . . . that seem to "cross-over" or to violate the radical separation of the spirit and unseen realms from the material world' (J. Mack, *Passport to the Cosmos: Human Transformation and Alien Encounters*, London: Thorsons, 2000, p. 25, cited in G. Hancock, *Supernatural: Meetings with the Ancient Teachers of Mankind*, London: Arrow, 2006, p. 379).

6 Personal communication (2009). See K. L. R. Jansen, 'Ketamine, Near-Birth and Near-Death Experiences', in *Ketamine: Dreams*

and Realities (Florida: Multidisciplinary Association for Psychedelic Studies, 2001), pp. 92–166.

7 Many cultures have similar myths: for instance, it appears in Russia as the story of the firebird and in India as the myth of Soma, Indra and the eagle.

8 R. G. Wasson, *Lecture to the Mycological Society of America*, 1961.

9 A. Huxley, 'Drugs that shape men's minds', *Saturday Evening Post*, 18 October 1958.

10 Ibid.

11 See, e.g., Lev. 19:31; 20:6; Deut. 18:10; cf. 2 Kgs 9:22; 21:6; 2 Chr. 33:6; Isa. 8:19; Mic. 5:12; Gal. 5:20.

12 Gen. 3:7.

13 Gen. 3:22.

14 Gen. 3:23–4.

15 C. Foster, *The Selfless Gene: Living with God and Darwin* (London: Hodder & Stoughton, 2009).

Chapter 13: Breathing God

1 J. Hitt, 'This is your brain on God', http://www.wired.com/wired/archive/7.11/persinger_pr.html.

2 K. Wilber, *One Taste* (Boston: Shambhala Publications, 1999), cited by J. Horgan, *Rational Mysticism* (Boston/New York: Houghton Mifflin, 2003), p. 58.

3 'The silence and the emptiness [in my heart] is so great that I look and do not see, listen and do not hear,' Mother Teresa told a spiritual adviser, and in a note written in 1955 she said, 'The more I want Him, the less I am wanted ... Such deep longing for God – and ... repulsed – empty – no faith – no love – no zeal.' She wrote a letter to Jesus: 'Lord, my God, who am I that You should forsake me? The child of Your love – and now become as the most hated one ... You have thrown away as unwanted – unloved ... So many unanswered questions live within me afraid to uncover them – because of the blasphemy – If there be a God

– please forgive me . . . I am told that God loves me, and yet the reality of darkness & coldness & emptiness is so great that nothing touches my soul.'

4 G. Household, *Against the Wind* (Boston/Toronto: Little, Brown, 1958), pp. 42–3.

5 C. S. Lewis, *The Problem of Pain* (London: Centenary, 1940), p. 141.

6 Cited in Horgan, *Rational Mysticism*, p. 117.

Epilogue

1 Notably, and famously, the New Caledonian crow. There is also a lot of work on scrub jays.

2 Re-run the evolutionary tape, said Stephen Jay Gould, and you'd have a completely different biology. See S. J. Gould, *Wonderful Life: The Burgess Shale and the Nature of History* (New York: WW Norton, 1989). But not so, it seems. Evolution appears to have a limited number of solutions available to it, or at least adopted by it. The consequence is that it is much more *directional* than is often thought. Far and away the most articulate and influential advocate of this thesis is Simon Conway Morris, *Life's Solution: Inevitable Humans in a Lonely Universe* (Cambridge: Cambridge University Press, 2003).

3 Explored in the Appendix.

Appendix

1 I conflate the terms too in the discussion that follows. Philosophers and psychologists distinguish for very good reasons between these terms, but the distinctions do not matter for present purposes. Professional philosophers might get upset by the slapdash way in which I have used 'self', 'consciousness' and 'soul' as near synonyms. Sorry. 'Consciousness' and 'self', for instance, are not necessarily the same, although there is a close relationship between

them. But it would be a ponderous book that referred consistently to 'the conscious perception of self' instead of 'I', in actual or implied inverted commas. I personally find confusing one of the traditionally used definitions of consciousness, framed in terms of the question, 'What is it like to be X?' It is presumably not 'like' anything to be a table, and accordingly a table is not conscious, goes the lecture. It is 'like' something to be the girl in the library, and she is therefore conscious. But can one sensibly ask, 'What is it like to be a bat?' And so on. This definition seems to me to import all sorts of more or less dubious assumptions about the relationship between sentience and subjective consciousness.

2 The left side of the world is affected rather than the right because of the cross-over of neurones. This is the phenomenon of 'hemifield neglect', discussed by Susan Blackmore in *Consciousness: A Very Short Introduction* (Oxford: Oxford University Press, 2006), pp. 24–6.

3 O. Sacks, *The Man who Mistook his Wife for a Hat* (London: Picador, 1986), ch. 2, 'The Lost Mariner'.

4 This example is discussed in Blackmore, *Consciousness*, pp. 36–9.

5 This view is more or less what philosophers, taking their cue from Derek Parfit, describe as 'bundle theory' – the idea that I am no more than the bundle of experiences that have happened to me. Parfit proposes that the Buddha was the first bundle theorist – an idea to which we return later. But even Buddhism is rather inconsistent. 'Look at the flags,' the teacher famously said. 'What do you see?' 'I see the wind moving the flags,' said one student. 'I see the flags moving the wind,' said another student, obviously thinking the teacher would be impressed by his cleverness. 'Neither of you is right,' said the teacher. 'It is your mind that moves.' But what is there to move?

6 His 'multiple drafts' theory.

7 This is known colloquially as 'Libet's delay'. Also, in the 1960s, Grey Warter implanted electrodes into patients' motor cortices

and used the amplified output from those electrodes to control a slide projector. The patients were alarmed, saying that the slide moved all by itself, just as they were about to move it. Discussed in Blackmore, *Consciousness*, pp. 86–8.

8 C. S. Soon, M. B. Brass, H.-J. Heinze and J.-D. Haynes, 'Unconscious determinants of free decisions in the human brain', *Nature Neuroscience*, 11, 2008, pp. 543–5. The abstract reads, 'There has been a long controversy as to whether subjectively "free" decisions are determined by brain activity ahead of time. We found that the outcome of a decision can be encoded in brain activity of prefrontal and parietal cortex up to 10 s before it enters awareness. This delay presumably reflects the operation of a network of high-level control areas that begin to prepare an upcoming decision long before it enters awareness.'

9 Blackmore, *Consciousness*, p. 86.

10 F. Crick, *The Astonishing Hypothesis: The Scientific Search for the Soul* (New York: Charles Scribner's Sons, 1994), p. 3.

11 Blackmore, *Consciousness*, p. 81.

12 R. L. Morris, S. B. Harary, J. Janis, J. Hartwell and W. G. Roll, 'Studies of communication during out-of-body experiences', *Journal of the Society for Psychical Research*, 72, 1978, pp. 1–22, cited in S. Blackmore, 'Near-Death Experiences: In or Out of the Body?', *Skeptical Inquirer*, 16, 1991, pp. 34–5.

13 A case can be made for the existence of Theory of Mind (which presumably at least correlates with consciousness in the sense that we are talking about it) in chimpanzees and other higher primates, and many would say that consciousness appears in many 'lower' animals too – notably, and controversially, the New Caledonian crow.

14 See C. Foster, *The Selfless Gene: Living with God and Darwin* (London: Hodder & Stoughton, 2009).

Index

(Numerals in italics denote references in the Notes)

Index

Index

health
  in relation to religious experience
    72–3
  of the world, impact of religion on
    289
hearing, last sense to go in death 158
Heinrich, Clark 118–19
Heissig, Walter 231
'hemifield neglect' 265–6
henbane 108
Henderson, Mark 41
Hesiod 66
Hicks, Bill 97
Higgins, Charlotte 65
Hildegard of Bingen 60–3, 222, 299,
    303
Hinduism 112, 131, 133, 155, 173, 192,
    226, 245
hippocampus 25, 35
Hitt, Jack 239
Hof, Wim 175
Hofmann, Albert 100
Holy Spirit 37, 88, 91, 121, 123, 240
  see also charismatic worship
Homer 65–6
Homo sapiens, evolution of 204–6
Horgan, John 56, 126–7, 276, 279–80
Horizons (BBC programme) 49
Horus 227, 228
Household, Geoffrey 133, 243
human–animal hybrids (therianthropes)
    101, 196–8, 207–9, 226, 228–30
hunger 74, 142, 174, 177
  see also ascetics
'hunting magic' cave painting theory
    208
Huxley, Aldous 24, 83–4, 102–3, 110,
    121, 131, 181, 235
hyperactive agent detection device
    (HADD) 191–3, 312
hypnotherapy 84–8
hypoxic brain 158, 240
hysterical conversion 79–80

'I', nature of 253–68, 318–19
iboga 101–2, 112, 235
Ibogaine 123
The Iliad 67, 260
incarnation, the 275–6

incense 7–8, 237
indifference, religious – in relation to
    genes 41–7
ineffability, sense of associated with
    NDEs 163–4
Inquisition, the 289
'Intimations of Immortality from
    Recollections of Early Childhood'
    (William Wordsworth) 25, 188
introspection, lack of 65–7
Inuit, the 91, 212, 226, 240, 282
Islam 131, 173, 192, 239

Jainism 133
James, William 82–3, 112, 184–5, 315
Jansen, Karl 127, 165–6, 233, 308
Jaynes, Julian 64–5, 66, 67, 286
Jennings, G. J. 282
Jeremiah 300
Jerome, Jerome K. 79–80, 290
Jerusalem syndrome 58, 67–71, 74,
    287–9
Jesus
  ascension of 91–2
  as crucified shaman 213
  see also Eucharist, the
Jews 67, 92, 116, 176, 242
  see also Judaism; Old Testament
Jimmie G. 257–8, 264–5
Joan of Arc 53, 223
John of the Cross, St 141–2, 235, 240,
    302–3
John the Baptist 79
John, of Patmos 53, 102, 119, 221–2,
    224, 292
Johns Hopkins University (2006
    psilocybin trial) 126–7, 300
Judaism 96, 176
  see also Jews; Old Testament
'junk' DNA 219, 314

Kalahari, bushmen of 31, 91, 247
Kalian, M. 288–9
kava 232, 235
Kempe, Margery 287–8
kensho 16, 277
ketamine 84, 111, 127, 147, 165–6,
    182–4, 233, 308
Kfar Shaul 67–9, 287

Index

morphine 16, 57

Morris, Simon Conway 195, *318*

mortification 121, 142, 235

Moses 219–20, 221, 239

MRI scans 2, 7–8, 29, 32, 87, *280*

multagnosia 277

'multiple drafts' theory *319*

mummification 146, 173

Murphy, Todd 57

mushroom cult 116–17

mushrooms, magic 101, 113, 115, 116, 119–20, 123, 186, 199, 217–18, 222, 232

  Amanita muscaria (fly agaric mushroom) 106–7, 113–15, 116, 117–19, *214*, 299

  psilocybin mushrooms 32, 97, 103–4, 106, 113, 118, 123–6, 217–18, 235, 293–4, *300–1*, *316*

Myers, P. Z. 44–6

mysticism 13, 43, 94–6, 131, 187, 201, 241–2

  mystical experiences 7, 15–16, 30, 38, 73, 82, 109–11, 120–1, 125–6, 128–9, 161, 182, 185–7, 201, 235–6, 239, 241–2

  mystics 18, 109, 121, 235, 245, 251

  see also altered states of consciousness; stigmata

myth 92, 106, 116, 187–8, 229, 260, 262, *316*

Native American Church 102, *316*

natural selection 46, 73, 187, 193, 199, 204, 225, 230, 266, *315*

NDEs (near-death experiences) 50, 89, 145, 148–66, 182–3, 233, 240, *305–8*

Neanderthals 202, 205

Nechung Oracle 178–9

Neolithic 214–15, *309*

Nephilim 221, 224

neuro-imaging studies 66, 129

New Caledonian Crow *320*

Newberg, Andrew 14–15, 26, 28, 30, 32–4, 36, 38–9, 129, 147, 187–8, 209, 217, 225, 250, 279–80, 281–2, 302, *310*, *315*

Nietzsche, Friedrich 39

Nirvana 175, 241

nitrous oxide 112, 146, 184–5

NMDA receptors 166

noradrenaline 43

North American Indians 113, 197

nuns, SPECT scan experiments on 26–7, 29, 32–4, 209, *281*

OBEs (out-of-body experiences) 1, 78, 145, 146–8, 154, 156, 158–63, 183, *304*

  see also NDEs (near-death experiences)

Obsessive Compulsive Disorder (OCD) 33

occult training 146

Odour of Sanctity 75

The Odyssey (Homer) 65, 67, 107

Old Testament, suggested psychoactive substances in 113, 114–15, 117–18

One Taste (Ken Wilber) 241

opiates, endogenous 16, 163, 174

opium 107–8, 110, 132, 245, 295

orgasm 15, 31, 133, 136, 143, 302

origins of religion, the 195–230

out-of-body experiences (OBEs) 1, 78, 145, 146–8, 154, 156, 158–63, 183, *304*

  see also NDEs (near-death experiences)

Pahnke, Walter 123, 126

pain 25, 74, 94, 147, 168, 174, 177–8

  see also masochism; pierced shaman

Palaeolithic, Upper 205

Panofsky, Erwin 299

parasympathetic systems 30–2

Parfit, Derek *319*

parietal lobes 28–30, 147, 187, 209, 244–6, 250, *281*, *310*

Paul, St 28, 37, 52–3, 68, 74, 78, 84, 139, 145, 222, 224, 245, 284, 292

peripheral awareness 87

Persephone 92, 99–100

Persinger, Michael 7, 13–14, 27, 54–7, 169–70, 233, 285

PET scan 11, 263

Peter, vision of in Joppa 291–2

petroglyphs 210

Index